Another Science is Possible

Another Science is Possible

A Manifesto for Slow Science

Isabelle Stengers
Translated by Stephen Muecke

polity

First published in French (excluding Chapter 4) as *Une autre science est possible!*
Manifeste pour un ralentissement des sciences © Éditions La Découverte, Paris, France,
2013

French text of Chapter 4 © Isabelle Stengers

This English edition © Polity Press, 2018

Polity Press
65 Bridge Street
Cambridge CB2 1UR, UK

Polity Press
101 Station Landing, Suite 300
Medford, MA 02155, USA

ISBN-13: 978-1-5095-2180-7
ISBN-13: 978-1-5095-2181-4 (pb)

A catalogue record for this book is available from the British Library.

Library of Congress Cataloging-in-Publication Data

Names: Stengers, Isabelle, author.
Title: Another science is possible : a manifesto for slow science / Isabelle
   Stengers.
Other titles: Autre science est possible! English
Description: English edition. | Cambridge, UK : Polity, [2017] | Includes
   bibliographical references.
Identifiers: LCCN 2017026397 (print) | LCCN 2017027589 (ebook) | ISBN
   9781509521838 (Mobi) | ISBN 9781509521845 (Epub) | ISBN 9781509521807
   (hardback) | ISBN 9781509521814 (pbk.)
Subjects: LCSH: Research--Social aspects. | Science--Social aspects. |
   Science--Philosophy.
Classification: LCC Q180.A1 (ebook) | LCC Q180.A1 S73513 2017 (print) | DDC
   501--dc23
LC record available at https://lccn.loc.gov/2017026397

Typeset in 11 on 14 pt Sabon by Servis Filmsetting Ltd, Stockport, Cheshire
Printed and bound in the United Kingdom by Clays Ltd, St Ives PLC

For further information on Polity, visit our website: politybooks.com

For the GECo.
For Serge Gutwirth
For all those who allowed me
to think that this is not simply utopian.

Contents

I

Towards a Public Intelligence of the Sciences

Should 'the public' 'understand' the sciences?

Our Anglophone friends speak of the 'public under-standing of science'.[1] But what is meant by 'understand' here? Many people think each citizen should have the basic 'scientific equipment' (or literacy) necessary to understand the world we live in, and especially to accept the legitimacy of the transformations of the world that the sciences bring about. In fact, when the public begins to resist an innovation that scientists have backed, as notably in the case of GMOs, the usual diagnosis points to the lack of such understanding. Thus, the public apparently fails to understand that the genetic modifi-cation of plants is not 'essentially' different from what farmers have been doing for millennia, but is just faster and more effective. Others say that the methods that make for 'scientificity' have to be understood first, and that the public supposedly mixes up 'facts' and 'values' because it doesn't understand that scientists are free not to ask certain questions. Of course, it is not a matter of denying citizens the right to accept or reject an innovation, but they should do so only on the basis of solid reasons, and not confuse scientific facts with their own convictions or values. The need for an apprenticeship in the sciences, it is argued, is founded on the fact that close observation, the formulation of hypotheses and their verification or refutation, form the basis not only for the construction of scientific knowledge but for all rational procedures. The sciences are therefore a model that every citizen should follow in their daily lives.

Such arguments are used today to justify a veritable 'order word'[2] coming from public authorities when faced with a somewhat suspicious citizenry. If the latter are sceptical about the benefits the sciences bring to society, the response will be: 'The public and its science have to understand each other.' The possessive 'its' implies what standard science lessons in school try to get across: scientific reasoning belongs by right to all, in the sense that, confronted with the same 'facts' as Galileo or Maxwell, each of us could have drawn the same conclusions.

Of course, anyone with even a minimal exposure to the history of science, or to the sciences themselves 'as they are made', can easily conclude that the anonymous rational being drawing these 'same conclusions' is just the correlate of the 'rational reconstruction' of the situation, from which any reason for hesitation has been purged, and where the facts literally 'shout out' the conclusion they lead to with all the authority one could wish for.

In any event, laboratory conditions, reconstructed or not, have very little to do with those situations we are confronted with as citizens. For the latter, I would use Bruno Latour's felicitous phrase, 'matters of concern', which, in opposition to what are presented as 'matters of fact', insists that we think, hesitate, imagine and take sides. 'Concern' happily incorporates the notions of pre-occupation and choice, but also the idea that there are situations that concern us before they become objects of preoccupation or choice, situations which, in order to be appropriately characterised, demand that 'we feel concerned'. We should not talk about these situations being 'politicised', as too many scientists complain. They are a long way from being occasions for the more or less arbitrary or contingent expression of political engagement; rather, what they require is the power to make people think about what concerns them, and to refuse any appeal to *'matters of fact'* that would bring about a consensus. If there is a question to be asked, then, it is first of all how such situations have so often come to be separated from this very same power, which they require.

To return to GMOs, they constitute a quite differ-ent 'matter of concern' from laboratory GMOs defined in terms of the preoccupations of biologists working away in well-monitored spaces. GMOs cultivated across thousands of hectares raise questions to do with genetic transfer and pesticide-resistant insects, questions that can't be raised at the level of the laboratory, not to mention issues such as patent applications for modified plants, the reduction (already critical) of biodiversity, or the runaway use of pesticides and fertilizer.

The essential thing with 'matters of concern' is to get rid of the idea that there is a single 'right answer' and instead to put what are often difficult choices on the table, necessitating a process of hesitation, concentration and attentive scrutiny – and this despite the complaints of the entrepreneurs, for whom time is money and who demand that everything that is not prohibited be allowed. Then there is the propaganda, often in conjunction with scientific expertise, that all too frequently presents an innovation as 'the' correct solution 'in the name of science'. This is why I would propose, in place of the notion of understanding, a 'public intelligence' [*intelligence publique*] of the sciences, involving the creation of intelligent relationships not just with scientific outcomes, but with scientists themselves.

What should the public understand?

When we speak of public intelligence, we have to emphasise first of all that it is not a matter of activists denouncing, as enemy number one, those biologists who have presented GMOs as 'the' rational and objective solution to the problem of world hunger. Rather, if a public intelligence is necessary, it essentially has to do with the very fact that those scientists were able to take this kind of position without a care in the world. If we put to one side hypotheses about dishonesty or conflicts of interest, then the question becomes one of understanding how the training and practice of researchers can lead to such arrogant and naive forms of communication, completely devoid of the critical thinking they so often boast about. How can one explain also the failure of the scientific community to publicly express outrage over this abuse of authority?

Quite the opposite occurred, it seems. Consider this extract from the summary report for the *États généraux de la recherche* held in 2004, in which researchers told the public what they should be understanding:

> Citizens expect solutions from science for all sorts of social problems: unemployment, depleted oil reserves, pollution, cancer . . . the path that leads to the answers to these questions is not as direct as a programmatic vision of research would have us believe . . . Science can only function by dealing with its own problems in its own way, shielded from urgency and from the distortions inherent in economic and social contingencies.[3]

This quotation comes from a collective report, not the wild imagination of some individual. Its authors not only attribute to citizens the belief that science can solve problems like unemployment, they too seem to agree with this belief. Apparently, science can solve problems like this, but only if it is allowed the freedom to formulate its own questions, shielded from the 'distortions' said to be 'inherent' in 'contingent' economic and social preoccupations. In other words, authentic scientific solutions transcend such contingencies, and thus can ignore them (just as those biologists cheerleading for GMOs have ignored the economic and social dimensions of world hunger).

In short, what I have dubbed 'matters of concern' are characterised as 'distortions' in this account, while the solution that 'science' comes up with is identified as an answer to a problem that has at last been well-formulated. It follows that citizens are right to be trusting, but they have to know how to wait, and understand that scientists owe it to themselves to remain deaf to any noisy or anxious demands.

In fact, in 2004, the researchers did not address

citizens, but went over their heads to the public authorities in charge of the politics of science, on the occasion of its redefinition in the terms of the 'knowledge economy'. In their complaint they took up the hackneyed theme of the goose that lays the golden egg – stand back, keep it well fed, and don't ask difficult questions, otherwise you will kill it and there will be no more eggs. Of course, it is not the business of the goose to wonder for whom her eggs are golden, and the generally beneficial character of scientific progress is taken for granted. The small question as to why this progress may today be associated with 'unsustainable development' is not asked.

I don't think that scientists are 'naive', like the goose whose egg we remove from under it in order to give it a new value for the sake of humankind. They know perfectly well how to attract the interest of those capable of turning their results into gold. But they also know that the knowledge economy marks the end of the compromise that guaranteed them a minimum of vital independence. They can't, however, talk about that openly, because they fear that if the public were to become aware of the ways in which science 'is made', they would lose confidence and reduce scientific proposals to simple expressions of particular interests. 'People' must continue to believe in the fable of 'free' research, driven by curiosity alone towards the discovery of the mysteries of the world (the kind of candy that helps so many well-meaning scientists to set about seducing childish souls).

In short, scientists have good reason to be uneasy, but they can't say so. They can no more denounce those who feed them than parents can argue in front of their children. Nothing should upset the confident belief in Science, nor should 'people' be urged to get involved in questions they are not, in any case, capable of understanding.

Sciences need connoisseurs

If public intelligence on scientific questions has any meaning, it is in relation to this type of systematic distancing. Scientific institutions, the State and industry all find their interests converging here. But we should not be naive about this either. We should not set up, in opposition to an infantile public in need of comfort, the figure of a thoughtful, reliable public capable of participating in the things that concern it. One initial way of not being naive is to remind oneself over and over, as the physicist Jean-Marc Lévy-Leblond has constantly done, that the question of being capable or not is equally relevant to scientists themselves. When he wrote 'If scientism and irrationalism, traditional foes, are still going strong, it is because uncultivated science turns as easily into the cult of science as into occult science',[4] he was not just talking about the public, but also, perhaps above all, about scientists themselves. In other words, a public intelligence of science would involve an intelligent and lucid relationship to scientific claims, an intelligence that would concern the scientists as much as the 'people', since they are all vulnerable to the same temptation.

We know that what Lévy-Leblond calls scientific culture is not to be confused with some general scientific literacy – knowing 'something' about physical laws, atoms, DNA, etc. A cultivated science should produce not only specialists but also connoisseurs, as is the case in sport, music or software production, i.e., in domains where producers know that they have to take into account the existence of people who are able to evaluate the products, assess the kind of information they are given, discuss its relevance, and differentiate between mere propaganda and calculated risk. For specialists, the existence of such connoisseurs, or amateurs, creates a demanding environment, which obliges them to maintain a 'cultivated' relationship with whatever they are proposing – they know the danger of skipping over the weak points, because the people they are addressing will pay just as much attention to whatever is neglected or omitted as to what is asserted.

So let's take up Lévy-Leblond's clarion call, 'There are no amateurs of science', because it throws new light on the question of the public intelligence of science. It is not a matter of asking the general question, 'Does the public have the capacity?', but one of asserting that it doesn't have the *means* to be capable. The 'indifferent confidence' of the public (that the scientists feel they have to protect against doubt) betrays above all the absence of demanding connoisseurs likely to hold scientists to the task of taking care when making normative judgements about what does or does not matter, or of presenting their results in a lucid manner that actively situates them in relation to the questions they really can answer, rather than as a response to whatever is the object of a more general interest. Had such an environment existed in 2004, the researchers would have thought twice before writing what they did.

It goes without saying that this is not a public where everyone would become a 'connoisseur' in every scientific field, a kind of generalised *'amatorat'*, or group of amateurs. But it could be a 'distributed amatorat', a multiplicity of connoisseurs dense enough so that those who aren't connoisseurs in a given field can be confident that should it ever concern them, they will be able to approach it in an intelligent fashion thanks to the milieu of connoisseurs which has already formed around it.

Let me emphasise that the 'connoisseur' here has nothing to do with the autodidact, especially the kind that scientists (and even a philosopher like me) are only too familiar with as those unfortunates who go around desperately trying to get recognition, or at least a little attention, for their solution to some major problem. Connoisseurs are not advocates of 'alternative' knowledge, looking for professional recognition. But their interest in the knowledges produced by scientists is different from the interest of the producers of these knowledges. It is for this reason that they can appreciate the originality or the relevance of an idea but also pay attention to questions or possibilities that were not taken into account in its production, but that might become important in other circumstances. In other words, they are able to play a role the crucial character of which must be recognised by all those who care about rationality. They are agents of resistance against a scientific knowledge that pretends it has general authority; they partake in the production of what Donna Haraway calls 'situated knowledges'.

Good will is not enough

These days, where the knowledge economy prevails, scientists could well have a vital need for a public intelligence nurtured by a milieu of connoisseurs. In the same way that uncultivated science can easily turn into occult science or into the cult of science, an indifferent confidence can tip towards mistrust or hostility. This is all the more likely to the extent that more organic links between research and private interests are being forged. Henceforth, those scientists who fight to conserve some basic autonomy will not be able to limit themselves to an appeal to 'save research'. They will have to have the courage to say what it is that research needs to be saved from; they will have to go public on the ways in which they are urged or compelled to become simple providers of industrial opportunities. And they will need a public intelligence that is inclined to hear them.

But the scientists will also have to know how to earn the support that they need, which will not be the case unless they are capable of hearing and taking seriously those questions and objections which today they too often dismiss as opinions that 'don't understand the science'. From this point of view, it seems to me disappointing and unsettling that agronomists, field biologists, specialists in population genetics and others, who were at first excluded from the commissions dealing with GMOs and their associated risks, did not loudly and clearly acknowledge their debt to those whose efforts had secured them some kind of a hearing in the first place – that is, those adversarial groups who were able to persuade the public authorities to adopt a slightly more lucid position on the GMO question, and who brought them more generally out into the political, social and scientific culture.

Here the scientific ethos itself is at stake, and in particular scientists' mistrust of everything that runs the risk of 'mixing up' what they consider to be 'facts' and

values'. This deeply embedded mistrust is quite different from a simple ignorance that might be remedied by
courses on epistemology or the history of science. My
teaching experience tells me that most students enrolled
in the so-called 'hard sciences' make up their minds to
forget such courses once they've got through the exams.
No surprises there, because by signing up for a 'hard
science' degree they have made a choice that is not initially motivated by 'curiosity' – or the 'desire to uncover
the mysteries of the universe' (most students arriving
with this in mind quickly realise their mistake) – but
by the image of the sciences promoted by the education
system. They have learned that the sciences allow problems to be 'well-posed' and therefore amenable to being
given the 'right solutions'. And those solutions will be
beyond dispute, verifiable by anyone, thereby silencing
those chatterers who mix everything up. However partial and deceitful this image may be, it has the power
to attract and select. Those who opt to take scientific
studies might be inclined to tolerate courses they consider 'mere talk', but they will not see them as a crucial
part of their training – and many of their 'real' teachers
will not fail to reinforce this prejudice with their shrugs,
ironic smiles and wise counsel on the importance of not
'spreading oneself too thinly'. Of course, any scientist
worthy of the name will be ready to swear allegiance
to epistemological principles concerning the limits of
knowledge and its conditions of validity, but only in a
formal way, because these principles will be forgotten
the moment a situation arises in which their knowledge
appears to be offering the 'correct', ultimately 'rational'
solution to some question that has exercised the chatterers. Clearly, this ethos implies that scientists refuse to

allow their own type of knowledge to be made part of the general culture: in their view amateurs are just chatterers who descend on these correct solutions and drag them into a world of idle gossip.

While it may be pointless to hope that courses on epistemology or the history of science could transform this situation, an experiment carried out over three years at the university in Brussels gave me a glimpse of another possibility.[5] A framework was set up within which science students were confronted with socio-technico-scientific controversies, but they were given sole responsibility for exploring the issues via the resources on the internet, and hence for discovering, in their own way and with no predetermined method, the clashing arguments, the partial and partisan truths, as well as the huge range of facts involved. Unlike other frameworks for the 'investigation of controversies' (in particular that of Bruno Latour with his students at Sciences Po in Paris), it was not a case of taking part in the construction of a new kind of expertise. The framework applied to any student, and its ambition went no further than that of complicating their 'thought habits'.

It became apparent that the students were interested in finding things out 'in the field', that is, on the Web: a field constituted by a variety of situations marked by uncertainty and by the entanglement of what they had assumed would be separable into 'facts' and 'values'. They had been in the habit of relegating to 'ethics' (no one speaks of politics any more) everything that didn't seem to bow down before the authority of the 'facts'. They discovered that there are many conflicting types of 'facts', and that each of them was linked, for those presenting them, to what appeared to be important in the situation. But they didn't draw sceptical or relativist conclusions from this discovery, because they realised that it was the situation itself (as a 'matter of concern') which imposed this conflictual entanglement that prohibited one order of importance (for example, that of proof) from dominating all the others. What indeed surprised them was the casual way in which scientists allow themselves to pooh-pooh as 'non-scientific' or 'ideological' things that others think are important.

I wouldn't say that these students were inoculated once and for all against the scientific rationality/mere opinion opposition, but I was impressed by the fact that, far from being plunged into chaos, confusion and doubt, at least some of them seemed to experience a feeling of liberation. It was as if they had discovered with relief that they didn't have to choose between facts and values, between their scientific loyalty and (the remains of) their social conscience, because it was the situation itself that required them to identify the relevance of a knowledge and to understand its selective character – what it makes important, what it neglects. It was as if this curiosity so often associated with science was being called upon and nourished for the first time.

Experiments like the one I have just described are obviously not sufficient, but they are perhaps necessary to weaken the hold of the slogans reproduced in such a remarkable way in the 2004 warning from the French research body. It seems that curiosity, much more than the critical reflexivity close to the hearts of epistemologists, is what needs to be nourished and freed from judgements about what does and doesn't count. Perhaps this curiosity could bring together students from different fields, allowing them to work together, collectively confronting situations that force them to take a distance from their respective favourite abstractions, and above all to overcome two fears: on the one hand, that of 'hard' scientists confronted with questions they 'can't deal with'; on the other, that of 'literary' or 'humanities' specialists faced with the authority of the so-called hard sciences. In short, working together, they could develop a taste for what I am calling 'intelligence'. And there will never be a public intelligence of science as long as scientists themselves don't have a taste for intelligence.

Science on trial

Scientists need such a public intelligence to emerge not only because they are faced with the now limitless power of their traditional industrial allies, but also because of another snowballing threat. I just gave the example of the resource-rich internet, but the internet is also, of course, a prime vehicle for rumours, conspiracy theories, and the most extravagant ideas. From this point of view, the picture-postcard image the sciences have given themselves is turned back against them, because extravagant ideas can endorse the same image, proposing 'facts' which should be conclusive but for the 'orthodox' scientists who refuse to take them into account being conformist, blind, timid, or even corrupt. Here we are paying a high price for the absence of a culture of 'facts' – of their rigorous fabrication through a laborious collective process by which 'viable facts' and the theories they authorise are co-constructed.

But this raises another question. Such a process is costly in terms of both work and resources, and scientists will only embark on it if it looks like being 'worth the trouble' as far as they (and their funding bodies) are concerned. Scientists are in general tight-lipped on the question of selection criteria, but, like the 2004 researchers, they feel that only scientists are capable of discerning promising lines of inquiry, thus claiming the right to ignore or exclude others. If necessary, they will limit themselves to justifying their research choices via a few arguments that are sometimes superficial and often presented with a somewhat dogmatic haste (refining the arguments would mean wasting valuable time).

The internet transforms this situation, however, because it makes it possible for a large audience to put forward counter-arguments that expose the weakness of the reasons given. And the counter-attack will be all the more formidable if it can draw on numerous cases of conflicts of interest, and denounce the way in which the science in question ignores facts which run against the interests it serves. While the reasons scientists have for not thinking a proposition worthy of attention may be trustworthy, such accusations persist because that trust can easily be devalued by the nature of the knowledge economy and the dependence it introduces between research choices and private interests.

The situation associated with the new public image of science that has taken hold – that of a dishonest and compromised business resisted only by a few valiant defenders of free truth – is a catastrophic one. Even more so in that scientists are ill-equipped to cope with it. They only have in-house communicators available, and a shortage of 'free' allies on the internet. They thus pay a heavy price for the absence of an 'intelligent' relation to the sciences, one that is *involved*, critical and demanding, and which is cultivated by 'connoisseurs', i.e. by those capable of hearing the reasons for the choices made, and of discussing and defending them if necessary.

But here once again, the support of such 'free' allies has to be earned. Their existence presupposes that scientists are prepared to give an account of their choices in a mode that doesn't insult the intelligence of the connoisseurs, that produces 'stuff to think with' that will nourish interesting debates, in short, that does not abandon the field to the myopic game of either attacking scientific authority or denouncing the 'rising tide of irrationality'. To the extent that the capacity to give such an account demands intelligence and imagination, it should be possible for the criteria that determine what is worthy of interest to become a little more open, a little less determined by conformity, fashionable priorities and hardened attitudes.

The current situation is all the more catastrophic in that the internet audience consists not only of busy isolated individuals, more or less enlightened and usually sincere, but also of shrewd, paid strategists. Naomi Oreskes and Erik M. Conway's passionate and disturbing book, *Merchants of Doubt*, reveals the long-term damage done by such 'merchants' in their attempts to undermine the credibility of scientific research bearing on 'inconvenient' problems, from the dangers of tobacco through to the threats posed by climate change today.[6]

Inconvenient truths

Ever since Galileo, scientists have congratulated themselves on coming up with 'inconvenient truths'. We perhaps take it as given that the Earth is not the centre of the universe, but it is not quite so straightforward with biological evolution, which, since Darwin, has 'inconvenienced' those who hold to the letter of the Bible (or the Koran). And yet there is a big difference between these believers and those who today pay the merchants of doubt to ensure the maximum publicity for their assertions. The evolutionary thesis inconveniences the believers because it contradicts the biblical account according to which each species was created separately. The 'truths' targeted by the merchants of doubt are inconvenient not because of what they contradict but because of their political and economic consequences. Scientists then discover, sometimes to their astonishment, that their traditional allies can be relied upon only when the 'facts' help 'increase productivity' – when that isn't the case, they are open to being transformed into promoters of unrelenting scepticism.

But there is a common thread between the believers and the merchants expressed in the sceptical refrain: 'There is no proof, so it is only an opinion, and can therefore be put on a par with other opinions.' The idea that it is the authority of proof that makes the difference between science and opinion is here turned against the scientists themselves.

That idea of authority has an incontestable relevance when we are dealing with experimental sciences, but when it is generalised to 'field' sciences, or to any situation that can't be sufficiently purified to render it testable and reproducible, a unified facade is created that is easy to knock down. This is why evolutionary scientists should recognise that what they call the 'proofs of biological evolution' are the kind of proofs that would make experimentalists chuckle. They should be brave enough to acknowledge that their facts are simply pointers. Then they would be free to give weight to what really matters to them, that is, to the way in which such facts, ever since Darwin, have been proliferating and making the story of life on Earth ever more dense and stimulating. As Stephen J. Gould has admirably demonstrated, what gives the evolutionary sciences their robust character is not the 'proof', but the number and variety of cases that become intelligible and interesting in a Darwinian perspective. This fecundity is perfectly sufficient to differentiate evolutionary theory from creationism and Intelligent Design, which are not characterised by any process of this kind since the author they put in charge is capable of explaining everything and anything.

The merchants of doubt also use the notion of 'scientific proof' to attack researchers who, while they do their best, are implicated in things that have nothing to do with experimental situations set up to respond to precise questions. Like anti-Darwinians, the merchants take advantage of debates among specialists – for whom it is quite normal to draw on models of interlocking processes as well as data from the field – to present the difficulties they deal with as crucial disagreements that 'they are hiding from us'. In the name of the 'balance of opinions' that has to be respected (since in the absence of a proof there is only opinion), the 'sceptics' demand that their case be heard whenever and wherever the question of climate change arises. And they have well and truly succeeded in creating the impression that the debate is still open, that the scientists really are divided, and that the dangers are perhaps exaggerated.

When presented as being founded on the authority of facts, the sciences had no need of connoisseurs.

Worse, they were suspicious of anyone who insisted a bit too forcefully on the irreducible plurality of scientific practices, that is, on the falsity of the image of a monotonous scientific progress leading to the triumph of a 'scientific reality' with answers for every question humans might ask. Today, the situation has changed, because the self-image of Science as the 'thinking head of humanity' has backfired on scientific institutions. This image was only good for generating respect; but it leaves science defenceless when it comes to confronting its real enemies.

Resisting the merchants of doubt

The history of life on Earth is fascinating, as evidenced particularly by the success of Stephen J. Gould's books. It can be appreciated by connoisseurs interested in the richness of the perspectives it opens up. In this sense, one can say that the best allies of the creationists are those leading lights who propagate the idea of evolutionary science as being intrinsically polemical, devoted to inconveniencing all who refuse to reduce their 'behaviour' to an effect of what would be the sole scientific explanation: selection. On the contrary, at the risk of shocking scientists, I don't think it so crucial that all inhabitants of the Earth accept the evolutionary perspective as quickly as possible. And it is from this double point of view that there are grounds for distinguishing anti-evolutionist doubt from what is currently peddled by the 'merchants of doubt'.

Quite obviously, these merchants are for the most part paid by industries whose interests are indeed 'inconvenienced'. But not all. Some are mobilised against anything that would inconvenience the grand narrative of human progress liberated by reason, or against the dangerous confusion of 'facts' and 'values' that an 'alarmist' science would foster, creating an alliance with critics of development and free enterprise. But, in the end, who amongst us would not wish for the prospect of climatic disorder to disappear? Who would not want the world to appear less dangerous, and our activities and lifestyles to have more benign consequences than they do? We are all vulnerable to the temptation to put our heads in the sand when confronted with this particular 'inconvenient truth'.

Furthermore, in this instance, time matters. We know this already with respect to climate change: as the Cassandras at the IPCC have warned us, the catastrophic could well become cataclysmic should we carry on as if nothing were happening, other than making a few cosmetic adjustments (we often forget that Cassandra *was right*). But it also matters for those industries claiming that, in the absence of certitude, more research is necessary, and that it is better to wait for definite proof. What they want us to forget is that if an incontestable certitude does arise, it will not have had a scientific origin, but will rather be a sign that too much time has been spent waiting and dithering, that 'reality' itself has decided to stage the demonstration, much to our annoyance. For those in business, gaining time doesn't just mean making money for a little bit longer, it also means preparing the way for a future in which we will have no other choice than to turn to them and their 'solutions', which can duly be presented as 'unfortunate but necessary'.

The question of public intelligence – as applied to the

plurality of sciences and what we can legitimately ask of each – may appear quite insignificant when faced with this kind of perspective. Yet ignoring this plurality, and continuing to promote the model of 'sciences which prove things', has given the merchants of doubt the capacity to attack with impunity. The scientists under 'attack' are not, as Oreskes and Conway show, 'heroes' capable of flamboyant counter-attacks, ready to publicly denounce the personal harassment they have been the victims of, vigorously demonstrating the dishonesty of their adversaries. This is not something they have been selected or trained for; on the contrary, they share the common scientific ethos that implies one should keep the public at a respectful distance, and that the sole authentic task of the scientist is to produce knowledge. Everything else, including the battle against deceitful representations of their work, is an unfortunate distraction and waste of time. What climate scientists in particular need is a public understanding of what it takes to decipher the climate, mediated by connoisseurs capable of mobilising against the strategies of their attackers.

Given that the future probably holds, more than ever, an increasing number of 'inconvenient truths', the question of public intelligence will tie science and politics together with a hitherto unseen intensity. How can one battle against the appropriation by scientists of 'matters of concern', of choices bearing on the common future, and at the same time learn to identify the 'merchants of doubt', disqualifying them in a public and merciless manner, as we have learned to do with historical negationists, promoters of racism and certain warmongers (*pace* Bernard-Henri[7])? How can one stop scientists under attack from making the opposition between science and opinion even more rigid than it already is? How also can one stop those with good reason to be wary of scientific claims to authority from giving in to the seductions of organised doubt?

Here, as elsewhere, time is running out. So it is worrying to remember that it is now thirty years since Jean-Marc Lévy-Leblond first raised the alarm, spelling out how unhealthy it is for science to be incapable of nourishing the milieus of connoisseurship that are still today sadly lacking.

2

Researchers With the Right Stuff

The gender of science

I would like to begin with what is undoubtedly the most common way of talking about the relationship between science and gender.[1] We all know that our political and scientific authorities are concerned about young people's lack of interest in the sciences. They are not concerned about history, sociology or psychology, but about those sciences that decision-makers in America refer to as *sound*. This means both sciences with a proven track record, and those that are capable of proving things. *Sound science* is a term that is even more impolite than 'hard science', because the opposite of *sound* (doubtful, suspect, fake) is frankly pejorative. Only sciences that can prove things, that is, that can evoke facts as authoritative, are worthy enough to avoid disqualification, and these are the sciences that young people are leaving in droves.

The idea that it is gender construction that keeps

---

women out of scientific research – even though they constitute a readily available human resource at a time of techno-scientific labour shortage – emerges in this context. Since it is foolish to neglect a part of the pool that the future of research and innovation will depend upon, it will be a matter of getting 'girls' interested in a career that they are supposed to be avoiding only because of a 'gendered' representation. In principle, it is claimed, science is equally open to all, and girls don't choose to study it only because they believe it isn't for them. In other words, the gender problem here relates only to an illusory representation, which better information, or a change of image, should be able to correct. The reality of the situation would be that science is neutral as far as gender goes.

The decrease in the number of young people embarking on scientific careers is often analysed as a social symptom. Today, apparently, young people refuse to dedicate themselves to the demanding requirements of the 'real' sciences, and would rather look for jobs with immediate payback. So the sciences are the innocent victims of a social fact. Critics argue the case that society no longer knows how to pay tribute to the great adventure carried out by researchers in the name of us all, and even that it is unfaithful to humanity's true vocation.

This talk of a true vocation, symbolised by curiosity, uncovering the mysteries of the universe, and the benefits brought about by scientific progress, might raise a smile. But it is this vocation that is advertised to young people, and especially very young ones. Thinking of the way in which scientific institutions try to encourage a taste for the sciences, one could almost speak – dare I say – of a kind of paedophilia, a thirst to capture the soul of the child. It associates science with a taste for strange gadgetry and disinterested questions, with the thirst for understanding and for science as a big adventure. Such tastes are, of course, no longer on the agenda by the time students enter university, and even less so when they start thinking about a research career. Far from being treated as a primary resource that is now under threat, young researchers of either gender, doctoral students or postdocs, have to accept the realities of onerous working conditions and fierce competition. They are supposed to grin and bear it: the great adventure of human curiosity presented to them as children is replaced by the theme of a vocation that demands body-and-soul commitment. And this is what we accuse today's young people of no longer accepting: compliance with the sacrifices that service to science demands.

What defines the scientific vocation, what stuff is a *real* researcher made of? It is clearly going to be a gendered construction in the sense that it has direct discriminatory effects for most women. One could say that the research career was designed for men, and even specifically for men who benefit from the support of women at home – bringing up children, taking care of practical matters, allowing them to do all-nighters at the laboratory and go off on numerous training workshops or on the kind of overseas trips expected in a research career. In the case of women, the price paid for such a career is all the more discriminatory in that judgements as to who counts as a 'real researcher' are part of the very definition of the vocation. It will often be said of a woman who also has family responsibilities that the very fact that she chose to take on such responsibilities shows that perhaps she never had the 'stuff' of a real researcher.

Heroic acceptance of such sacrifices is the way in which vocation is proven from the start. If a researcher, male or female, gives up, they will be said not to have had 'the right stuff' – as in the book by Tom Wolfe,[2] which tells the story of test pilots training to become the first astronauts in NASA's Project Mercury. If a test pilot died on the job, his colleagues would say that 'he didn't have the right stuff'. The interesting thing was that there was no positive definition of this stuff, given there were multiple reasons why a pilot could be killed, mostly depending on the plane he was testing. It is precisely this unacceptable degree of dependency that the expression hides: whatever flying coffin they were given to test, those who were killed didn't have the right stuff.

It should not be necessary to point out that this question of 'stuff', in the way I'm beginning to talk about it, is not directly related to research skills. No one says that the pilots who died were bad pilots. Rather, the talk of stuff indirectly indicates what can never be questioned, something no one will speak about or blame: the technical viability of the prototypes the pilots had to test. So there is something a little more subtle going on here than is captured by familiar sociological categories such as ideal type or habitus. The question of stuff points fairly specifically to the construction of a difference linked to questions *that are on the table, but will not be asked*, to a kind of gritting of the teeth and resistance to what is an ever-present temptation. Test pilots have to resist expressing concern about what is for them a matter of life and death; the test pilot gets behind the controls of whatever plane he is given, no questions asked.

This is all about what the pilot is worth, in the sense of Boltanski and Thévenot's *economies of worth*, where they discuss judgements about what is deemed 'great' and what 'petty'.[3] However, the 'stuff' that a test pilot is made of, his worth, seems to me to have the essential feature of a 'gendered' worth because, in contrast with Boltanski and Thévenot's sense, it is defined in the negative: the 'real pilot' is the non-marked standard, just as 'men' are in regard to women. We don't know what makes for a good pilot. Those who are marked are those who are killed. Only the crash, therefore, is witness to what they didn't have but the others possess. One could speak of the mysteries of divine election here, but neither scientists as a group nor test pilots seem to me to be inhabited by this kind of mystery. The kind of construction we are dealing with is unique in that it doesn't pretend to describe a reality, so it would be in vain to call it illusory. Rather it is 'true' in the sense that it makes things 'hang together', creating a particular relationship of self and other. It both presupposes and produces an ethos.

It is this ethos, this stuff, that I want to focus on here. It is a construction whose prototype is certainly the differentiation of men and women, but which also spreads everywhere. As it happens, the construction of a 'real test pilot' is confined to an exclusively virile group, while the duty of the widows, and of the wives of those who survive, is to keep quiet, to not make a fuss.

Real researchers

On the basis of this hypothesis, to interrogate the stuff that makes for a 'real researcher' (including the women considered worthy of this title) is to interrogate a construction gifted with a remarkable power because it doesn't deform reality, but requires a determined insensitivity to the questions posed by that reality. These are usually expressed in a mode of denial: 'we know, of course, but anyway . . .', and in any case real researchers have to grit their teeth and ignore such questions.

There is no doubt that in certain countries (notoriously, not in France), feminism has raised new questions about the way knowledges are cultivated in our academic worlds, and has challenged numerous aspects of the scientific ethos. But today another notable feminist thinker is reasserting her relevance: Virginia Woolf (I believe I can hear her laughing sarcastically). Her book *Three Guineas* consists of three interlinked replies to three demands to join causes, each time consensually.[4] The responses are tough, with a painful lucidity that forces us not to think about consensus as a gesture of good will. It is not too difficult to imagine what Woolf would have thought about contemporary appeals to 'save research'.[5] This has nothing to do with declaring null and void the attempts of past feminists to bring about 'another science'. Hearing Woolf laugh gives one more of a sense of the distance separating us from a time when her thought might have been considered too pessimistic, when it was still thought that the brutal academic ethos was something that girls would fail to transform, and that therefore they should avoid joining the ranks of the long procession of 'cultivated men'. As much as this procession has, today, lost most of its magnificence, and is looking a bit shabby and uneasy, it still excludes men and women who insist that it stop, just for a moment, to think: to take the time to ask the question that Woolf insisted we should never stop asking. 'Let us never cease from thinking', she wrote, thinking of all times and places, 'what is this "civilization" in which we find ourselves?'[6] And, by extension, what is this academic world that is being destroyed in the name of excellence? We have to think in order not to fall into the trap of a nostalgia for a world which is actually in the process of collapsing into the past.

Woolf's analysis of this world in *Three Guineas* is unremittingly tough. Of course, she resists the temptation to take petrol and matches to the prestigious English colleges that churn out people who are both conformist and secretly violent, whose capacity for violence flares up when they sense they are in danger. She resists only because it is still there that girls can obtain the diplomas that will enable them to make a living. But they should avoid making their careers in such institutions, or in any other profession promising prestige and influence. They should use the university to acquire knowledge that actually emancipates them, but they should remain outsiders. Otherwise they will have to conform to the ethos that such professions demand: aggressive competition, intellectual prostitution and an attachment to abstract ideals.

In short, I think that Virginia Woolf had an excellent fix on what I am calling the 'researcher's stuff', and I don't think she would be at all surprised at the submission and passivity of today's academics in the face of the redefinition of their world and their practices. This redefinition is being carried out in the name of objectively evaluated excellence, in a mode that well and truly demands the systematic practice of the kind of intellectual prostitution Woolf denounced. Because not only does this stuff *not* characterise a 'good' researcher, determining only what constitutes a 'real' one, it could well have something to do with the terrible transformation Woolf describes when the 'private brother, whom many of us have reason to respect', is swallowed up and replaced by 'a monstrous male, loud of voice, hard of fist, childishly intent upon scoring the floor of the earth with chalk marks, within whose mystic boundaries human beings are penned, rigidly, separately, artificially. . .'.[7] This brutal and puerile male often appears when he feels that the 'mystic demarcation' separating 'real scientists' from other humans is under threat or being 'relativised'; when he feels that the way in which most scientists present themselves, and are represented – that is, as heroic researchers resisting the temptations of 'opinion' – is in danger. This violent being is also manipulable, precisely because this demarcation is abstract, lacking any content apart from its opposition to this marked 'other' that they call 'opinion'. Those who 'don't want to know about' anything that might cause them to hesitate are always manipulable.

Scientists, they say, have objectivity as their common 'worth', and this might actually be the sole claim that could bring together practices as diverse as physics, sociology, psychology or history. And yet, it is remarkable that all attempts by epistemologists to identify the content that would unite these different practices have run up against banalities devoid of any relevance. In fact, I would venture to suggest that the only thing that can bring them together is nothing other than the definition of opinion as irrational, subjective, malleable and condemned to illusion and appearances. This informs, by the way, the content that Gaston Bachelard attributes to scientific rationality: an ascetic 'no' directed towards a veritable gallery of horrible opinions. As Bachelard puts it: 'By right, opinion is always wrong, even in cases when, in fact, it is right.' This is the *cri du coeur* of the 'real researcher', his 'I don't want to know about it.' The test pilot 'wants to know nothing' about the criteria that differentiate the plane he is about to test from a flying coffin. The real researcher wants to know nothing about a world in which sometimes 'opinion is right'.

Let's face it, today the majority of scientific expertise is tasked with keeping the anxieties of opinion under control, letting it know it is wrong and that it is incapable of the objective judgement which is the privilege of scientists. And it is because this is a real duty, agreed to in the name of the general interest, that the relevance of such expertise is rarely discussed at the heart of the academic world. The objective point of view held by the expert has to be (and often it is enough that it is) in stark contrast with the subjectivity of questions that are important for 'opinion'.

And yet those who have to make decisions complain about scientific expertise because it is too hesitant for their taste – it weighs up the issue for and against, muddying the water when what is wanted is a definite answer 'in the name of science'. The 'worth' of the decision-maker (another unmarked type) lies in knowing that a line must be drawn. And he would like the experts to tell him where that line should be drawn: 'Be men; not finicky, chattering sissies. If you mean yes, say yes! Don't wallow in doubt or incertitude.'

What is this objectivity that we are on a mission to defend? Because the only general answer to this question mobilises 'facts' capable of relegating whatever divides opinion to subjectivity, it is easy enough for those who know how to spin the key phrases to trap scientists and make them toe the line. If 'facts' are opposed to values, and are capable of rendering any question 'objectively decidable', how can one resist the directive to make this capacity prevail? When scientists have replied 'present!' to the call to be ready to decide on any hesitant issue, this fraudulence has not generally been denounced by their colleagues. Those who have decided that in order to keep opinion quiet it is necessary to present a united front, i.e. a 'scientific method' assuring objectivity, have had to tolerate the proliferation of experts armed with new methods whose characteristic blindness has become synonymous with objectivity. The 'data-' or 'evidence-based' sciences have given themselves the project of defining any situation or choice in terms that allow objectively measurable data to evaluate and decide the issue.

Here too we are dealing with a real ethos, a mission that mobilises real crusaders, who will dismiss the debates and hesitations of their colleagues as simple expressions of opinion from those ignorant of the fact that the only well-put questions are those over which the facts can adjudicate. The circle closes when 'excellence' – the new slogan applying as much to universities as to research groups and individual researchers – is benchmarked against such data. It was scientists who constructed such methods with impunity, and their colleagues failed to challenge them even as the same methods were used against them. Today they are discovering the consequences in an increasingly direct manner.

As we know, with such data-based evaluations it is not a matter of taking the particularities of each university into account, still less of finding out about the work of individual researchers. That would run the risk of troubling judgement, of returning to hesitation. Data are objective in the sense that they are 'unmarked', and so capable of being used as benchmarks against which all is measured, without hesitation or discussion.

Today we encounter everywhere this 'gendered stuff' which defines the worth against which people without the stuff discuss, think and hesitate – this stuff which has nothing to say for itself, except that it must be accepted in the name of what Woolf identified so well as abstract, mystical ideals. And as she diagnosed it, these ideals are inseparable from brutal disqualification and noisy advertising; and from the stupid pride of resisting the insistent question that women must ask themselves over and over, everywhere and always: what is this civilisation in which we find ourselves?

The construction of a real researcher

Thinking along these lines implies resisting nostalgia. No doubt things were better in the past, but what is happening now is logical enough, and that logic was already at work in the past. This is what I would like to develop by doing a little history, not of the sciences, but of this researcher's 'stuff', this ethos that claims to be synonymous with the spirit of science, and that has ended up today with a definition of excellence 'based on the facts'. My aim is not to act as an historian, but to sharpen the appetite for possibilities that run the risk of being obscured by denunciations of the present in the name of a past that we can always idealise.

My starting point is the work of Elizabeth Potter,[8] whose importance was highlighted by Donna Haraway in her book *Modest Witness*.[9] Potter shows that gender was well and truly in play in the modes of experimental life that Robert Boyle endeavoured to promote. More specifically, she pinpoints the question of gender as a difficulty that could cause the whole experiment to collapse.

In fact, how can one uphold a man's virile qualities if he does not heroically risk his life, cultivate personal glory, or allow himself to be carried away by his passions or his opinions? How can one talk of the virility of a man who presents himself as a modest witness, deferring to the facts and seeking no glory other than that of revealing them? Isn't the reputation of the gentleman engaged in the experimental life in danger if he claims the modesty and reserve usually expected of the feminine gender? Aren't such chaste beings going to be disqualified for lack of virile virtues if they refuse the joys of flamboyant rhetorical conquests?

But chastity and modesty are not the lot of women alone; they also define the correct disposition for the service of God. What Boyle proposed was the worth of spiritual (not corporeal) chastity and modesty, a discipline of monastic origins. He who follows the experimental path serves God via the disciplined exercise of reason. And this reason is well and truly virile in the sense that it is part of masculine heroism to make an abstraction of one's own interests, of one's prejudices, and to resist the temptations and seductions of questions that would lead one astray.

I have personally witnessed the power of this construction, the way it is able to ensure that the disciplinary order rules supreme. It happened when I was a chemistry student. I excluded myself from future research because I thought I had irremediably strayed from the path. The question of whether I had the researcher's stuff was not asked – as in the case of the test pilots, judgement is retroactive, coming only after the accident. In my case it came after I let myself become interested in what scientists call the 'big questions', the so-called non-scientific questions.

However, there is a distinction to be drawn between Boyle's chaste and modest researcher and what led me to conclude I was 'lost for science'. Boyle's researcher, if he gave into temptation, could repent, but I considered my self-disqualification as a researcher to be irreversible. Another type of ethos defining the real researcher comes in here. Dating back to the nineteenth century, it can be conveyed by way of the image of a sleepwalker who must not be woken. I was still conforming to this image when I realised that, since I had woken up, I should leave.

The sleepwalker is always perched on the ridge of a high roof, walking up and down without vertigo, fear or hesitation. He poses no questions that might throw him off balance. Chastity in the service of knowledge has been replaced by a sort of anthropology of creativity, with a thesis according to which the researcher must have a faith that will 'move mountains', that is, must not let his way be blocked by any obstacle to his quest for intelligibility – especially when these obstacles have already been gloriously dismissed as what 'opinion believes' before 'real science' intervenes. This faith often makes itself explicit in the negative: if one takes this dimension of the problem seriously, then science will not be possible. And it regularly ends up confirming the relevance of the 'lamp-post parable', in which a passer-by, stopping to help someone desperately looking for his keys at the foot of a lamp-post in the middle of the night, ends up asking: 'Are you sure this is where you lost them?' To which the other replies, 'Not at all, but this is the only well-lit spot!'

Having the right stuff, then, means having faith that what a scientific question doesn't make count, doesn't count; a faith that defines itself against doubt. He or she who has been bitten by doubt will not recover the faith that research requires. Waking up the sleepwalker kills the researcher.

Boyle's experimental scientist was chaste, and avoided any attachment to theological or metaphysical questions. The ethos of sleepwalker scientist, on the other hand, is more of a phobia. He rejects any questions he considers 'non-scientific' in a manner which is not without parallel to the phobic misogyny of the priesthood, meaning that he endows them with a dangerous, seductive power that is liable to lead him down the one-way road to perdition. Furthermore, the range of these questions has become broader, since they now encompass, for example, questions about the role of the sciences in society. Certainly, such questions can't be officially banished in the same way as theological and metaphysical ones can. But they are still half-implicitly dismissed through the subtle smile, the ill-disguised warning, or the snickering and gossip about so-and-so 'who doesn't do science any more'. Along the way, enemies will be made of those who insist that scientists ask themselves certain questions, or who demand that they give an account of precisely what it is they are defending in the name of science. Sleepwalkers refuse to hesitate when it is a matter of differentiating between what is important for them and what they judge to be secondary or anecdotal. Give us leave to be obstinate and aggressive, to decipher the world in terms of conquests and obstacles to be overcome, otherwise you will no longer have any researchers left! Such is the argument that those who advocate a new approach to science training have come up against.

For my part, I have stopped believing in the virtues of courses on the history of science, or on the social role of the sciences, at least as they are currently delivered to science students of either sex. Because every student enrolled in the ('hard') sciences knows perfectly well that these courses 'are not science', that as soon as the exam formalities are complete they will not really count. Most, in relation to these courses, are like the scientists invited to Diotima's receptions in Robert Musil's *The Man without Qualities*: smiling into their beards when confronted by men of learning.[10] The students listen politely to what they recognise to be big ideas, but they already know that 'real scientists' would never let themselves become infected by such things.

Such subtle smiles, rooted in this phobia, are a natural feature of the sciences that today's young people are deserting, much to the consternation of our governing bodies. It is these sciences that Thomas Kuhn identified in *The Structure of Scientific Revolutions* as functioning paradigmatically, and which he begins by characterising in terms of the question of how students are trained. Training in sociology or psychology involves a panorama of rival schools, courses in different methodologies, divergent definitions and debates, as students are introduced to the founding texts in their discipline, those that set out the choices that will engage them. In contrast, Kuhn emphasises, the strength of the paradigm is its invisibility. The young people being trained

37

are well and truly on track to become sleepwalkers for whom the right way to ask a question goes without saying: it relates to incontestable evidence. From this educational perspective, for a (hard) science student to read anything other than her textbooks is not only a waste of time; it is also a disturbing sign, a bad omen for her future, implying that she might not have the right stuff.

Boyle's chaste researcher has a general enough definition of the proper value of scientific objectivity: it requires a refusal of the 'big questions' that would seduce opinion, which is 'always wrong'. And this chastity can be claimed by all sciences, in the name of not confusing 'facts' and 'values'. For his part, however, the phobic sleepwalker belongs specifically to those sciences which, since the nineteenth century, may be characterised by their crucial role in the development of the so-called productive forces. And this is no accident. Sleepwalking researchers were born in a laboratory which is no longer analogous with the monastic discipline of the cultivation of the spirit, wherein wasting time was a sin. The laboratory is now defined by the imperatives of gaining time, competition and speed. It is no longer for the sake of ascetic discipline that researchers refrain from asking 'big questions', but rather because their training actively turns them away from such questions. Everything that might distance them from their discipline has been excluded, deemed a 'waste of time' or, worse, a pathway to doubt. In other words, the phobic, for whom doubt is the enemy, is first of all the person who has never learned to take a step sideways, and who therefore does not know how to slow down without losing their balance.

But, for all that, 'real' sleepwalking researchers are not totally blind to the world around them. They don't ignore it, but they certainly won't allow it the power to make them hesitate. They decipher it in terms of opportunities. One could even depict them as being on the alert, attentive to the possibilities of presenting what counts for them in a way that will interest whoever is likely to add value to their results. And they will be all the more innovative and free to be entrepreneurial to the extent that they despise, with a properly virile contempt, the multiple and interlocking aspects of the problem they are supposed to be looking into.

A recent and striking example is, of course, the claim of molecular biologists that their strains of genetically modified plants could solve the problem of world hunger. The gendered dimension was clear in the phobic contempt with which they dismissed the doubts of their colleagues who pointed to the socioeconomic reasons for famine, to social inequalities that were in danger of widening, to the destruction of agricultural modes of production, or to the difference between laboratory-created GMOs and those planted on hundreds of thousands of hectares. In this case, the social scientists and field scientists were like women with too many sensitivities, who can speak only of risks and uncertainties. Had we had listened to them in the past, we would have thought electricity dangerous, and we'd all still be getting around on horses and carts. A real researcher must know how to take risks on board and accept the price of progress. But as far as knowing who might be exposed to these risks, well that's a big question . . .

Let's not wait with too much confidence for the phobic sleepwalkers to 'wake up' to the damage done by the knowledge economy. One could say that, in different ways, researchers have been told that 'the party is over' – today they have to submit like everybody else to the same law. No one can walk away from the demand that flexibility and competition prevail everywhere. And that means the elimination, from any science whatsoever, of all those individuals who don't have or do what it takes to maintain a career. The brutal redefinition of their jobs has no doubt made plenty of researchers grumble, but, at the end of the day, not too loudly. And in tragi-comic fashion, many of them blamed 'opinion' (yes, once again) for its failure to understand that science has to be left alone to be fruitful. The politicians, infected by opinion, have ratified the 'rise of irrationality' that means the 'public' no longer respects science (hence the mass desertion of young people from scientific studies). The idea that there could be the slightest relationship between this defection and what is happening in the world seems almost unutterable. The advance of knowledge owes it to itself to persevere heroically through all kinds of hostilities.

We can predict that the next generation of researchers will smile cynically at evocations of the good old days when scientists asked their own questions. But a new gendered construction will certainly bless them for their courage in making common cause with the entrepreneurs, while more sensitive souls denounce ecological ravages and growing social inequalities. The 'real researcher' will be the one who knows that human destiny demands terrible sacrifices, and that nothing should hinder it. Meanwhile the new construction will only prolong the hatred, already cultivated in the name of progress, towards those chatterers with the big ideas spreading doubt, worry and disorder.

Ever since I began to get a clearer sense both of what was happening and of the relative submission and passivity of researchers, I have taken seriously what Virginia Woolf had already diagnosed as intellectual prostitution – the docility of those who, without being tied down like wage-labourers, nevertheless agree to work and think just where and how they are told. But in reality, where else can they turn when they have consistently opposed scientific objectivity to political preoccupations? How can they publicly discuss the issue of disaster when they don't want the public to lose confidence in 'its' science or start meddling in something that doesn't concern it? The researcher's 'right stuff', and his dependence on mystical demarcations, prohibit him from asking, together with others, Woolf's question about this civilisation in which we find ourselves. He can only groan and try – but every man for himself – to find ways and means to pursue what he will call 'good research' which 'advances science'.

Demobilisation?

Facile hopes are not possible when thinking with Virginia Woolf. Taking seriously a gendered construction like that of the 'real researcher' clarifies the violence that she describes all the way through *Three Guineas*: the violence of those who have learned to grit their teeth in order to stay the course, despite the sirens of temptation. The unmarked gender is equally defined by anxiety – the anxiety of being found not to have had the right stuff.

Besides, it is apparently the case that the first women primatologists – who didn't have this worry, having no hope of a career path – invented a 'slow primatology', which was not normalised by advancing the difference between matters that should interest a scientist and the seductions of opinion. They allowed themselves to be affected by the beings with whom they were dealing, looking for suitable relationships with them, putting the adventure of shared relevance above the authority of judgement. Their research reminds us that the way the researcher's stuff has been characterised is obviously not sufficient to define those research practices that make us willing, despite everything, to defend the university. The researcher's stuff no more makes someone into a researcher than the test pilot's stuff makes them a good pilot. The women primatologists offer an example of a research practice where the initial difference was connected with the fact that they weren't 'mobilised', i.e. summoned to prove they had the stuff of the 'real researcher'.

It is useful to remember that mobilisation is an affair of men at war. A mobilised army will not slow down for anything. The only question that matters is, 'can we get through?', and the price that others will pay for their passing though (ravaged fields, devastated villages) will cause no hesitation. Hesitation and scruples become synonymous with treason. Of course, scientists who hesitate are not executed, but the submission of the majority to the mantras that supposedly define the real researcher is sufficient to establish disciplinary mobilisation, because those who ask disqualified 'non-scientific' questions will always be in a minority, looked on with suspicion. People will question whether they can still be real researchers if they have let themselves be seduced by what a real researcher must keep at bay. In contrast, mobilised researchers will fall back on the quasi-automatic consensus of slogans like 'save research', without ever addressing the question, 'save it from what?'

So hope does not come easily; but I would like to introduce an unknown factor into the situation, and let the idea of the *possibility of demobilisation* resonate. The unknown factor is a gendered one, and this time it has a definitely marked gender, since women have always been suspected of being seductive or corrupting, inciting honest and courageous men to treason or desertion.[11] This unknown factor takes on a concrete meaning today, that is, a political one. My conviction is that the only possibility of 'saving research' goes by way of waking up the sleepwalkers, who will only wake up if they are compelled to do so. And they will not be compelled except by demands that rejig the question of what can or should be expected of researchers, by new requirements prohibiting them from adopting an attitude of denial when faced with questions that 'real' researchers have not been supposed to ask themselves.

Today such requirements are prefigured in what are called 'citizen juries', 'citizen consultations' or 'citizen conventions' – the term preferred by the French *Sciences Citoyennes* association.[12] Such panels, *when effective*, are meant to resist the set of catchwords and judgements that hierarchise different points of view. They constitute genuine tools [*dispositifs*] that level the playing field, resisting the scenario of: 'if you want to discuss things, you first have to stop being ignorant'. It is the jury that asks the questions, demands explanations, and evaluates the relevance of the answers it is given in relation to the problem with which it is concerned. It is the jury that summons counter-experts, that listens to the objections and organises confrontations. In short, it creates the type of testing-ground essential for the reliable evaluation of an innovation, because the concern for reliability excludes *a priori* any hierarchy between what counts and what may be overlooked, between what corresponds to an objective or scientific point of view and that which would be merely a matter of opinion or conviction.

The question of the role of such testing-grounds is a political matter, which means that the question of the making of researchers is a political matter. Their (very hypothetical) extension and generalisation would seriously test the duplicitous game typical of sleepwalking scientists: pretending to a humble ignorance of the 'big questions' – that is, of questions that do not interest their science – while presenting a problematic situation in such a way that whatever doesn't interest them appears as secondary, with the scientific point of view then appearing as the first, objective, rational step to take in approaching the problem.

The test will disqualify the sleepwalker, but it doesn't require scientists to take on board questions they don't know anything about, only that they actively situate what they do know – that is, explain how their knowledge can contribute to the problem, without identifying it with a 'scientific' or 'rational' perspective that would predetermine the way in which the problem should be cast. This seems like a fairly legitimate test, but researchers, at least as they are trained today, are very often not up to it, simply because it is difficult to situate oneself in relation to what one has been taught to despise – or at least to hold at a distance.

It is not a matter of an appeal for science to discover a conscience, or for researchers to take responsibility for the consequences of the innovations in which their research has participated. Nor is it a matter of opposing 'good science', serving genuine collective interests, to science biased by its service of private interests. In both cases, scientific knowledge still arrogates the crucial position of serving an interest that transcends particular passions. The test that interests me, corresponding to what Donna Haraway called 'situated knowledge' way back in 1988, designates that which precisely, and in a concrete fashion, has the task of questioning this privileged relationship of the sciences to questions of collective interest.[13]

Situating oneself has nothing to do with the Google Earth point of view, where you can see the whole Earth, then locate your own country, town, street and house. Being capable of situating oneself – situating what one knows, and actively linking it to questions that one brings in and to ways of working that respond to it – implies being indebted to the existence of others who ask different questions, importing them into the situation differently, relating to the situation in a way that resists appropriation in the name of any kind of abstract ideal.

Of course, citizen juries are rare and precarious, and what is more, easily gutted of any meaning. Since they presuppose a genuine political questioning of the sciences in their multiple and varied relations to innovation, they merely amuse those thinking in *Realpolitik* terms, which today means reducing politics to (good) governance. Their interest (the unknown factor with which I associate them) is to propose a distinction between scientific practices and the gendered construction that constitutes the stuff of the researcher. Citizen juries are carriers of a perspective that can contribute to breaking the impression of a fatality bearing down upon us where the role of science is concerned. Introducing an unknown factor has nothing to do with providing a ready-made solution, but entails casting a problem in such a way that its solution becomes conceivable. A solution exists, but it is not reached by way of a society that respects its researchers. It goes via a society that forces its researchers not to despise it.

In *Gender and Boyle's Law of Gases*, Elizabeth Potter relates how the high-society ladies who joined the audience for the air-pump experiments were upset at seeing asphyxiated birds suffering simply in order to prove that the air evacuated by the pump was necessary for life. Such a narrative can be associated with the long exclusion of women – they were unwelcome in laboratories – but it can also have another meaning that speaks to the possibility of a future wherein scientists will not smile into their beards upon hearing about such displays of feminine sensibility. There is no guarantee that in this future birds will no longer be sacrificed. On the other hand, the possibility of the scientist not smiling means that they will no longer cultivate their phobic fear that the questions and interests of others might demobilise them or cause them to lose precious time. They will have stopped pretending to the role of being the thinking brain of humanity and will have learned, through others and thanks to others, to appreciate the particularity of the questions that are important to them, questions that would then be stripped of their power to redefine or judge those that are of concern to others.

It is this 'thanks to others' that is important here. The unknown gendered factor in the question has no meaning outside of the perspective of struggle. But we are dealing here with a type of struggle that is in profound affinity with what women have been, and still are, fighting for: a society in which no single position can legitimate the silencing of others, who are supposed not to count. But it is also a struggle in which humour, laughter and mockery are crucial in face of the power of abstract ideals. Some researchers may learn to laugh at those who condemn them as traitors if they dare not devote everything to the advancement of science, avoiding idle questions. Demobilised, they will learn to appreciate the landscape that situates them, instead of passing through it at top speed.

3
Sciences and Values: How Can we Slow Down?

In the grip of evaluation

Today, publicly financed research is in the process of losing its autonomy. Researchers feel that they have been 'betrayed' by the political authorities, who, instead of respecting a consensually recognised right, have given corporations the power to select who among them will benefit from public sponsorship in every field where economic competition is in play. And where this isn't the case, where neither patent, nor partnership, nor 'spin off' are likely, a governing pseudo-market law has been put in place that is supposed to guarantee that public money will be used in the same kind of optimal fashion that the market, they say, would provide. The definition of the mechanisms of evaluation that are presented as 'objective', because they are blind to what counts for the researchers themselves, is an integral part of this enterprise. When the law of the market prevails, different actors, in competition with each other, have to be sensitive to 'signals' and respond with the greatest flexibility to the changing definitions of 'demand'. Where the market cannot be defined in terms of economic transactions, and where the definition of supply and demand is somewhat fictive, the mechanism of evaluation will have to activate this fiction. It will have to put those 'evaluated' into competition with each other in such a way that what matters for them, what makes their activity meaningful, ends up being defined as a 'rigidity', as something they will have to give up if they want to demonstrate their capacity to adapt.

As it happens, in the field of research, competition for the recognition of 'excellence', which has become a condition for academic survival, puts into play that rare resource which is publication in a top-ranking journal. This condition for publication thus demands that researchers pitch their research on the basis of what these journals impose in terms of norms: conformity, opportunism and flexibility – such is the formula for excellence.

People will say that I am overstating the case, that scientists know how to adapt to these new constraints without losing their creativity. And they will insist that these constraints at least have the clear advantage of weeding out the lazy, or those getting by quietly in some field that no one is interested in. But wherever this brand of 'new public management', as it is called, takes hold, the same story is repeated. It begins with consensual propositions, highlighting the advantages, and especially the increased 'transparency', as developments to be feared only by those seeking to 'profit from the system'; the others have nothing to fear, and indeed the formal nature of the evaluation should reassure them –

it isn't about controlling what they do. But then, those undergoing evaluation suddenly discover that the criteria applied – formal as they are, and blind to content – still contradict the sense of their work, and that they are non-negotiable. To begin with, they try to outsmart the criteria by cheating, but little by little the vice tightens. And at the end of the day they find themselves in a radically transformed landscape. Under constant surveillance and pressure, they have effectively been cut off from whatever it was they cared about. They are either reduced to the sadness we call depression, or end up as the kind of opportunistic cynics who know how to make all the right moves.

The ranking of specialised journals plays a key role in bringing researchers to heel. They discover that they must avoid publishing in minor journals dedicated to their type of research, and instead 'have to' publish in major journals, the criteria of which thus determine the value of their research. Before commenting upon this situation, I'd like to emphasise how insular specialised scientific journals are anyway, where articles are submitted to 'referees' chosen from among peers, that is, 'competent colleagues', and then read, in general, only by such colleagues. This peculiarity stems directly from the functioning of the 'modern sciences' themselves, where evaluation is immanent to the community, a community in which authors are read by other authors who assume the key role of taking into account, extending or contesting what they have read.

This traditional mode of evaluation should not be idealised. It has not stood up well to the explosion in the number of researchers and publications – 'publish or perish' did not arrive yesterday – nor to the increasingly rigorous linking of such forms of evaluation to tenure-track job selection. Furthermore, the referee system has been in a sorry state for some time; what had been a privileged responsibility has become a burdensome task done in a hurry, or perhaps an opportunity to get back at someone, or make career moves, or pass judgement on a reputation (blind reviewing does not mean authors can't be 'located'). As for 'collegial competence', it has become too fragmented to deliver on the evaluation of job candidates or research funding. Evaluation has gone down the path of bibliometric calculation, measuring the 'value' of an article by its citation count. Such procedures don't just offer ways of measuring the impact of a publication, useful for 'incompetent' evaluation committees. By decoupling evaluation from the competence of colleagues, who know how to judge the importance of a contribution in their own field, they have opened up the game to strategies such as click-counting or systematic mutual citation, against which defensive strategies have had to be developed in a kind of arms race that reminds one well and truly of Darwinian evolution.

In other words, the methods of evaluation now being imposed are not an attack on a system that was previously working satisfactorily. They have more to do with transforming the pressure to publish into a rigid imperative, a pressure that used to be deplored as an unfortunate tendency with related perverse effects. Those effects are now exploding. Without even mentioning fraud or misconduct, the number of articles 'withdrawn' after publication (meaning: 'should never have been accepted by the referees') is sharply increasing, including and even mostly in the top journals!

So it is easy to understand that, for those still attached

to research quality, one of the first ways to push back, apart from contesting the ranking of journals, is to slow down the number of publications and to insist that referees take the time to judge whether an argument is well-made or if it represents only a partial result, without intrinsic interest, hastily published to get a few points. However, I would like to go further. Even if the peer-review system worked perfectly – good articles being given the time to mature, referees being attentive and competent, etc. – it would remain the case that the various sciences, all the different ways of 'doing science', are not, never have been, and never will be, equal under this model of evaluation.

What I want to demonstrate here is that this model has been invented for the 'fast' sciences, with their strict differentiation between the cumulative production of knowledge addressed only to competent colleagues, and 'vulgarised' forms of knowledge. In conjunction with this, I would like to make a plea for a slowing down of the sciences. This would not be a return to a somewhat idealised past, where honest and worthy researchers were justly recognised by their peers. Rather, it should involve an active taking into account of the plurality of the sciences, in dialogue with a plural, negotiated and pragmatic (that is, evaluated on its effects) definition of the modes of evaluation and valorisation relevant to different types of research.

Who are the peers?

'Peers', or competent colleagues, and speed are two sides of the same coin. Both are translations of what makes possible a quite particular type of success, the success proper to the experimental sciences. This doesn't mean that successful experimentation was bound to be correlated with a fast science model – with competent colleagues as the only ones who can, or should, be allowed to evaluate – but that it is in relation to the experimental sciences that the model makes sense.

In order to characterise this success on the basis of its quite specific conditions (contrasted with the generality of abstraction as 'method'), I want to talk about it in terms of transplantation.[1] What is studied must be susceptible to extraction from one milieu and transplantation to another, typically that of the experimental laboratory. Only under this condition can 'experimental success' be achieved, because only in the laboratory can the questions posed receive so-called 'objective' answers, the publication of which is destined to be read by 'competent colleagues'; that is to say, by those who know how to evaluate them because they share not only the same milieu as the authors (their know-how and instruments), but also the same requirements when it comes to determining what counts as an 'objective answer', i.e. the same definition of the 'facts' deemed capable of authorising well-determined interpretations. The evaluation is therefore 'fast', not in the sense that it demands little work or effort, but in the sense that no objection will lead to compromise on questions of principle or doctrine, since it will correspond to the verification of the concerns of all the 'competents' involved – the extension of the domain of success. Do 'the facts' hold water? Do they authorise the author to conclude whatever it is she concludes?

This is why the researcher, as Bruno Latour has emphasised, is never alone in his laboratory; virtually present are all those whose objections can, and should, be anticipated. On the other hand, all those questions that transplantation excludes are absent. This is why addressing readers who belong to milieus in which other questions are cultivated poses a problem, one that is very often translated into an operation of capture.

Capture can happen in a great variety of ways, according to the ability of those-to-be-captured to frame their own conditions. At one extreme, there is industry, with its researchers working in over-equipped laboratories, and with lawyers, marketing teams, etc. The projected capture of its interest implies a consequential transformation of the scientific proposal, with a mass of grey literature, most often protected as commercial-in-confidence. At the other extreme, there is the 'general public', to whom scientists of good will – devoting part of their precious time to this charity work – will explain how 'science' is now capable of responding to their preoccupations and the questions they want to ask, including even those that Man has asked himself from the very beginning. These two types of transformation have very little in common, except that they don't retain what unites the researchers, what for them is important and which gives its own value to a new proposition: its cohort of 'but then . . .'; 'and therefore it should . . .'; 'what if. . .'. Industry will transform the 'new' into the 'innovative', while the general public will hear about a breakthrough that concerns the whole of humanity ('we believed, now we know. . .').

This broad-brush picture is both a little too caricatural and too indulgent. It is indulgent because with the knowledge economy – which might be better named 'the speculative economy of promises' – the distinctions are muddied. Faced with the fantastic promises coming, for example, from biotechnology, one is sometimes reminded of Neverland, where the pirates chase Peter Pan and the Lost Children, who are chased by the Indians, who are themselves chased by the wild animals who are chased by the Lost Children. Who believes whom, follows whom, is captured by the dream of whom? In the end, it doesn't matter anymore because the machine, in Félix Guattari's sense, now makes speculation and production coincide. It functions. Bubbles emerge and then burst, absorbing ever more capital, researchers and dreams. And the picture is a caricature because there are researchers-authors-critics for whom 'leaving the laboratory' deserves to be thought through with the same scrutiny as occurs inside the laboratory. Let's just say that they are not only in the minority, but are also viewed by their colleagues with a kind of suspicion, as if their loyalty towards the one thing that matters – the pure 'advancement' of knowledge – is in doubt. Besides, in a certain way, this suspicion is justified to the extent that what the example of these suspects shows is the non-contradiction between 'being situated' by belonging to a scientific collective and 'situating oneself' actively, that is creating relations with others who are not seeking some form of capture.

Let us now leave the experimental sciences, where the fast science model was invented, in order to pause for a moment at the opposite extreme, with the production of a knowledge which is not science: philosophy. And let us take the case of a well-known philosopher, Gilles Deleuze. How would he be evaluated? His citation count in the top philosophy journals (generally of the analytical persuasion) would be weak. As for his productivity, it would be judged derisory because he did not publish many articles, and most that he did publish appeared in journals that don't count. As for his books, they don't count either – a book is 'outside evaluation' because a 'real researcher' publishes for his colleagues, within the bounds imposed by referees. Fast evaluation 'by peers' thus condemns Deleuze's way of doing philosophy. On the other hand, there are philosophers who publish (only) for their colleagues, with abundant cross-referencing as they discuss, critique, complicate, complete and modify each other's arguments. We should not try to reconcile the modes of recognition and evaluation philosophy requires: for Deleuze himself, the academic prosperity of the 'fast philosophers' was coterminous with the assassination of philosophy.

But the question here is not one of opposing 'science' to philosophy. Rather, it cuts across all the disciplines, even as they are all officially subject to the same ideal model, that of judgement by 'competent colleagues' capable of evaluating the contribution of one of their own to the collective advancement of knowledge. In order to clarify this question, we have to choose a feature that can define this domain. Here I will choose to define the sciences via the specificity of a given collective work, where the value of an individual proposition lies in its 'contribution' to the collective dynamics. This in order to ask what a contribution is, as it effectively links competent colleagues together.

Certain fields, such as the neurosciences, are characterised by the rapidity with which they pile up publications showing all the signs of 'successful' laboratory work, with 'facts that demonstrate' at every turn. And some of these 'showings' have significant media repercussions in

the mode of: 'we had believed, but now we know. . .'.
But what seems much less common in such fields is the
type of dynamic that links 'competent colleagues' – as
indicated by references to works on which a particular
author's own claims depend, signifying a cumulative
dynamics wherein the recognised viability of a conclusion makes new questions possible. A good number of
neuroscientific demonstrations contribute only to an
accumulation of the kinds of 'facts' that are of no use
to working colleagues – even if they are candy for the
media. And in this case what links together competent colleagues could well be a kind of pact relating to
hypotheses that 'we all have to make' in order to confer
a definite meaning on what can be observed by way
of sophisticated instrumentation. To challenge these
hypotheses 'without which science would be impossible'
is as dangerous as violating a taboo: 'Don't touch that,
don't ask that question, otherwise we are no longer
scientific!' And this is how a mountain of 'methodologically impeccable' articles, can, as happened with
behavioural psychology, fall into insignificance when
something that was previously taboo now 'obviously'
has to be taken into account (even if it means creating
new taboos. . .).

In other fields, the notion of 'competent colleagues'
fails to unify because it runs up against divisions over
doctrine and conflicting ways of owning 'science's' her-
itage, even including the very definition of what passes
as a 'contribution'. These divisions are not simple com-
partmentalisations, but divisions among schools, each
often defined by an adjective indicating a founding
father – an adjective that marks both loyalty and a failure to eliminate rivals (Durkheim, Bourdieu, Chomsky,

or . . . had the ambition to reign without rivals, to usurp, like Newton or Lavoisier, the position of founder for the next scientific re-institution of their science). In these fields, the very idea of being evaluated by a colleague belonging to another school, or even of citing them in a reference, is meaningless, and each school has to 'possess' a top-rank journal as a matter of life and death.

These examples are no doubt extreme, but they have the virtue of centring the problem of differences among the sciences around the question of the collegial links that make for the novelty of so-called modern sciences. This question, by the way, is muted by the famous difference between so-called 'hard' and 'soft' sciences, a difference that brings in values of humanism, and the irreducibility of human relations to objective explanation, or to quantitative measures. The problem for 'soft' science is that it is on the defensive, and as such incapable of creating a positively different way of 'doing science' with its proper collective dynamic. This is why every time a conquering discipline moves forward – announcing that, at last, the really 'hard' science will rout the 'soft conversationalists' with 'truly objective' factual assaults – it does not generate an organised counter-offensive. The protests will too often be generalised and based on principle. Such an appeal to principle won't stop the conqueror from being immediately embraced as the representative of irreversible progress, and the rather summary presuppositions weaponised in its conquest will remain unchallenged. Rather than the refrain, 'ask a stupid question, get a stupid answer', however relevant it may often be, what triumphs is the refrain, 'physics, too, began with simplicity, with Galileo's falling bodies', intoned by those who rush to relegate the 'soft' to 'values' from which, we all know, 'real' science must dissociate itself.

'Science', an amalgam to be dissolved

Instituting a plurality of sciences against the unity of 'Science' means treating this unity as an amalgam that has to be dissolved in order to free the different ingredients in their particularity. Dissolving an amalgam doesn't involve passing judgement, but rather doing away with pseudo-similarities. For example, the authority of 'facts' in the sense of indicating a successful experiment certainly has nothing to do with the kind of authority in play when a toxicology test concludes that a certain product is of no danger to the public, or a clinical test determines that a certain drug can be granted medical status, put on the market and prescribed. In the first experimental case, the success is a kind of event, no doubt expected, but without guarantee. In the second, the conclusion follows a codified procedure that carries the guarantee of an answer within it. It is a matter not of judging the facts produced by such procedures, but of emphasising that they relate to a type of practice very different from that which produces 'experimental facts'. Even if what is submitted to the procedure has come out of a research laboratory, and even if the procedure itself calls for sophisticated instrumentation, the question it has to respond to is a question of public interest, and the authority clothed in facts is the fruit of a public decision.

Clinical and toxicological tests are not conducted according to a 'scientific, at last' definition of the therapeutic efficacy of a drug or the dangerousness of a product. They are conducted in response to the perfectly respectable necessity for legal or regulatory classification, even if this is a function of criteria that could be challenged on the basis of empirical data, as is the case today with endocrine disruptors. Here we can speak about 'conventions', or agreements negotiated among parties with conflicting interests; there is nothing dishonourable about this, but it requires particular attention and vigilance. Respect for such a convention requires keeping an eye on those who might abuse it for profit, or even cheat on the tests it involves. In this context, any argument that enters the fray appealing to the inherent authority of 'sciences that prove things' is a sign that one of the parties is up to no good.

In order to characterise these conventions, I would like to mobilise a type of scientific practice that is foreign to the standard notion of modern science; namely, that of the 'cameral sciences', defined by their service to the State in its role as guardian of public order and prosperity.[2] I think it is pertinent to expand these cameral sciences to include an ensemble of practices, whether laboratory work, statistical inquiries, or operational models that are used to make a decision (or where one hopes a decision will be reached). Such practices can no doubt be presented in terms of objectivity, method and facts, but what they produce must be understood as 'information' on a state of affairs, on a situation whose categories respond first and foremost to a power of acting, evaluating and regulating something exterior to them. One could say that they act like an organ of perception, selecting and giving form to whatever interests, or should interest, any institution having the power to relate consequences to perceptions. This formalisation could be called 'objectivation', a unilateral definition relative to the possibility of action.

Many sociological works, including critical ones, can therefore be classed on the side of cameral practices, and many experts from different scientific communities collaborate in them. It is quite obviously not a matter of criticising these practices, but of emphasising that they belong to a line of descent much older than that of the so-called modern sciences. They relate to the needs of any public or private 'government' – to the art of governance, and not to the creation of situations that allow, perhaps, new things to be learned. Those with an interest in what these practices produce should (ideally) be those whose actions are likely to be 'informed' by the knowledge produced. 'Peers' or 'competent colleagues' have no particular role to play here. On the other hand, defining what is relevant for these practices gives political action a very specific role. As Dewey showed in *The Public and Its Problems*,[3] as the GMO affair illustrated, and as the intervention of Act Up in the clinical testing of AIDS therapies demonstrates, political action aims to create a 'public issue'. It calls on state or state-like institutions to take on board new responsibilities or to modify their definition of public order, and therefore also how they define the information they need. Creating an issue is, for good or ill, quite properly a political event.

The question of the plurality of the sciences can only be raised after this first amalgam is dissolved, when the argument 'you have to accept this hypothesis, otherwise we can no longer define our object in a scientific manner' is referred to the imperative of objectivation proper to the cameral sciences. Then the expression 'in a scientific manner' is replaced by 'in a manner that makes a decision or an action possible'. And then comes the question of plurality: if we turn to those sciences that, unlike the cameral sciences, can be said to be 'modern', how can we dissolve a second amalgam produced, this time, by the injunction to establish 'facts' authorising an interpretation that will be said to be 'objective'?

The term 'objectivity', one might suspect, will not do because it facilitates all kinds of amalgams: between the object defined by the experimental sciences and the imperative to objectivation in the cameral sciences; between facts defined methodologically and experimental facts; between 'science' and what is opposed to it: irrational, subjective, egoist, etc., opinion. On the other hand, the question of success could well be connected to what links competent colleagues, what primordially matters for them as competent, and what it is that situates their competence.

There is something quite strange about the way the experimental sciences succeed. It is actually not enough just to extract the thing to be studied and transplant it into a milieu defined by the scientist's question. It is also necessary that this double operation not actively intervene in the type of response obtained, as is notably the case in pseudo-experimental situations where what is investigated is not only staged but it is anticipated to behave in a way that satisfies criteria of objectivity ('behaving like a rat'). What competent experimental colleagues worry about is that their extraction be well and truly 'purified' of any parasitic effects that might muddy the response to the question. This signifies in turn that the question is 'right'; that it addresses a dimension of the phenomenon that can be easily 'sorted out', and can therefore be attributed to the phenomenon independently of its milieu.

It should be clear from this that the conditions of experimental success are very restrictive, and from three points of view: Can what is studied be submitted to laboratory conditions? Can what the extraction eliminates be defined as simply 'parasitic' on the question? And finally, is what is being investigated indifferent to the intentionality inherent in the milieu to which it is transplanted, a milieu 'made' to get an answer from it? Is it 'its behaviour' that constitutes the response, or is this behaviour merely the way it replies *to* the scientist? This last condition dissolves the amalgam that was facilitated by terms like obedience and submission. Public enemy number one of experimental success corresponds to something the social sciences never exclude: the possibility that the 'subjects' may behave in the way they think the scientist is expecting them to behave.

From this last point of view, we can sketch another way of speaking about the contrast between the so-called 'hard' and 'soft' sciences. *A priori*, the questions put by 'hard' science only interest competent colleagues – hence, by the way, the need to solicit the interest of the 'public' (vulgarisation) and of those who can draw 'non-scientific' consequences ('benefits') from their propositions. A science will be called 'soft' when non-specialists feel competent to comment on it, to give their opinions on the questions it asks, because these questions concern or interest them. Hence the three ways of taking a distance from opinion: the cameral inquiries, the critical examination devoted to undermining 'opinions', and the submission of the object to a method that ensures the production of a 'different' knowledge which will interest only those for whom the primary measure of scientific progress is the way in which it triumphs over opinion.

In emphasising the extremely rigid character of what is needed for experimental success, I do not wish to confirm the privilege that the 'hard' experimental sciences already enjoy. The aim, rather, is to make room for other kinds of success that might prolong experimental success by reinventing it and associating it with other types of conditions. Such conditions don't have to be soft – they can be just as demanding as experimental conditions – but they 'simply' demand something quite different.

A perspective that can be called 'pragmatic' could then be substituted for the notion of a 'scientific vision' of the world conceived on the model of what is demanded by experimental success: a world fundamentally indifferent, certainly complicated, but proposing only one type of success, i.e. the discovery of the 'right point of view' that allows the 'right questions' to be asked, on the basis of which the jumble of empirical observations become intelligible. As long as this vision predominates, astronomy will be the authoritative precedent, because, they tell us, there was nothing but an accumulation of empirical facts until Kepler, followed by Newton, discovered the point of view that made them intelligible. So let's accumulate, and wait for the geniuses – a proposal which, when it appears in the neurophysiological literature, neglects the not-inconsequential difference between the heavens – which let themselves be observed with no questions asked about how the observation itself might disturb them – and a brain, the activity of which can only be studied if the subject endowed with the brain 'obeys' the experimental injunctions. A pragmatic approach would, on the contrary, give much more attention to this difference, which implies that the conditions of experimental success should be called into question.

Pragma means 'affair', and the affairs of scientists always proceed by putting things into relation, creating very particular relationships with other beings, so these beings have to answer a well-defined question. But there are many types of relation in this genre, including those under the headings of seduction, or torture, or statistical inquiry. . . . In the case of what I am calling the 'modern sciences', we are dealing with collective practices that assemble 'competent colleagues' around the question of the kinds of relationships that will allow them to *learn from* what they are studying. In other words, such relationships, in order to have a 'scientific' value that will prolong the values of the successful experiment, must allow what is being investigated to have the capacity to put at risk the question that is being asked of it.

This proposition is intended only to contribute to the opening of a problem, not the resolving of it. Because 'prolong', here, doesn't mean resemble. And 'having the capacity' doesn't just mean 'having the possibility of', but rather, when dealing with those 'polite' beings known as humans, that they should feel themselves empowered to understand – and, as the case may be, to contest – the way in which a question 'targets' them. This is why Bruno Latour, apropos the social sciences, suggested that critical social scientists committed a *felix culpa*, a mistake with fortunate consequences, when they took as their target practitioners of experimental sciences who protested violently.[4] They felt insulted when asked questions that didn't take into account what mattered to them: their success in conferring on their facts the power to make them all agree. For Latour, the (non-cameral) social sciences should take the lesson on board: they were at fault not only in this case, but also every time the people they study reply 'without complaining', whatever insult might be implied by the question. Only with 'recalcitrant' protagonists – those who demand that what matters for them be recognised and taken into account in how they are addressed – can a relation be created that has a claim to scientific value.

Contrasts

The typical risk for the experimental sciences is the accidental production of an 'artefact' for which the questioning operation can be shown to be partly responsible. This risk implies that what is questioned should be indifferent to the question it answers. In contrast, the social sciences as I characterise them require their subject's non-indifference to the question. Of course, this doesn't mean that those questioned have the right to dictate to the researchers how they want to be described; it means only that they have the capacity to evaluate the relevance of the relationship that is proposed to them. This first contrast activates others. So, it is certain that what a 'Latourian sociologist' reports to her colleagues will be quite different from what an experimental scientist reports to theirs, and this in at least three ways. First, she cannot claim to be dealing with facts that impose their own interpretation, thereby constituting her colleagues as verifiers who, in their own laboratories, are supposed to put to the test the consequences that 'should' follow (*doveria*, the first word in experimentation, inscribed by Galileo on the famous 116f folio[5]) or that 'could' follow ('but then' being the second word in experimentation). Second, her colleagues will no longer come together via a collective dynamic in which each successful establishment of a relation opens or closes new possibilities for other relations. Finally, they will come together even less in that, when it comes to the publication of results, they will not be the only addressees. In fact, a success of this type is likely to interest many people and, as the case may be, transform the way in which sociologists engage with and are assessed by other groups.

Here we are dealing with what is used as an argument for the 'soft' sciences: the difference between humans and the spheres rolling along Galileo's inclined plane that confirmed his *doveria*. And it is true that scientific practices that try to get around this difference are literally haunted by it: really spooked by the possibility that their subjects might understand how they 'should' respond. Particularly in experimental psychology, the interest that subjects take in the knowledge produced about them becomes a real curse, since what is being studied should be a 'behaviour' that is indifferent to the meaning of the question asked of it. Unlike those of magicians, the stratagems used to 'trick' the subject are not sufficiently secret or robust to prevent the 'facts' from being highly fragile, their lifecycle correlating with the plausibility of the naive credulity attributed to the subject.

And yet a contrast, rather than an opposition, should respond to this difference, a contrast bearing both upon the production of relations and the risks they involve, and upon the competent colleagues and the concern that links them. This is important because, without collegial links, the treasured notions of reflexivity and critical lucidity will have no effect. 'Soft' will remain soft, that is to say, deprived of the collective dynamic of knowledge-building that characterises the modern sciences. One might say that it doesn't matter, and perhaps it doesn't in some world other than ours. But in this world, where academic institutions have taken the research practices of the fast sciences and their competent colleagues as exemplary, anyone mimicking those sciences will always be at an advantage. It goes without saying that objective evaluation is dedicated to transforming this advantage into hegemony, pure and simple.

'Slowing down' the sciences is not itself the answer to the question of how to create contrasts among them, but it is the *sine qua non* for such an answer, and equally for the evaluative practices that would link colleagues in a manner free from the model of cumulative knowl-edge about a world considered as given. Our worlds demand other types of imagination than the 'so then that should . . .' or the 'but then that perhaps could. . .'. And corresponding to this plurality, there may be a plural dynamics of collective apprenticeship, putting into play what is meant, for each science, by the risk of establishing a relation.

I will take as a promising case the way in which cer-tain ethnologists learned to deploy what such relations were asking for when they took the risk of casting off the colonial anchors that had assured a stable difference between the ethnologist and those he was interrogating. What they reported back was less knowledge 'of' than knowledge 'between', knowledge indissociable from the transformation of the researcher herself, whose ques-tions were put to the test by other ways of making things, beings and relations matter. And it is to the extent that this type of transformation, with its risks and even dangers, concerns them all that her colleagues are 'competent', that is to say, primarily interested in what one of their number has learned, in the limits she came up against, and the way in which she was able to negotiate them or recognise their meaning, but also in the way in which she was forced to situate herself, to accept that her way of thinking, listening and anticipating situated her. This is what Eduardo Vivieros de Castro calls a 'decolonisation of thought' process, but my approach leads me to think of it without any connotations of blame or heroism, and rather in terms of apprenticeship – the ethnologist can certainly keep at the forefront of her mind the dense relation between ethnology and colonialism, but this is not what will make her capable of learning from those who agree to welcome her in.

Other fields offer us examples of somewhat similar collective, although less taxing, apprenticeships. These, notably, address what the researchers see as having archival status. Not just texts, but everything that is likely to be a witness to the past – the human past or the past of the Earth and its inhabitants. No doubt one can say that the archive is a 'given', even if what comes to be considered as an archive doesn't stop multiplying. But this very multiplication, the subtle interlocking of disparate remnants witnessing and taking their consistency the one from the other, contributes not only to more knowledge, but also to apprenticeships in new ways of narrating various pasts, of exploring their proper consistency, without submitting them to the kinds of simplifications defined by a 'progressivist' perspective using terms like 'still' or 'already'.

But it is above all the way in which the 'science' amalgam enters into conflict with what makes a science fruitful that we can feel other values pushing forward, values other than that of 'facts that prove', introducing other ways of evaluating. From this point of view, the field of evolutionary biology is remarkable. Since Darwin, it has been based on a refusal of the idea of progress leading towards the human, but it is haunted by a polemical pride in having thus illustrated a major trope of 'the science that debunks illusions'. As elsewhere, questions of how to 'narrate well' are multiplied, refined and dialogical, but nowhere else are they as stifled by a machine that reduces all of history to 'facts' that monotonously witness to the same truth: natural selection. But it is not just the histories told by evolutionary biologists that become 'matters of proof'.[6] From ethology to the human sciences, 'real science' publishes in prestigious journals 'facts' that are extracted in a brutal way from their contexts and interpreted as attesting (without, of course, the least 'it should' or 'but then . . .') to the general explanatory power of selection, set against the illusions of their 'backward' colleagues who are 'still' looking for other ways of learning. Nowhere else has the value of 'facts that prove' released so much destructive violence, supported by a model of evaluation that is deaf to the cries of those who see their field being ravaged by stupidity. Alas, poor Darwin![7]

The situation is a little different in field ethology. One could say that primatology set the example for an explicit apprenticeship trajectory, one pointedly celebrated by those who took part in it. It concerned the demanding character of the relationships that conferred on those under investigation the capacity to test the relevance of the questions being put to them. After a few years, primates, and later an increasingly large body of animals, escaped the status of being regarded as matters to be proven. Even where ethology is defined by a method that assures its 'scientificity', the norms censuring everything that might be suspected of anthropomorphism have lost their stability.[8] If a prestigious team dares to take seriously a question that had previously been laughed at, and a well-known journal publishes their results, this is now sufficient to lift the taboo, and teams of researchers rush through the open door. But even then the taboo remains intact in principle. The fact that what was once excluded is now included is celebrated as 'progress', and in no way compromises 'the method', outside of which everything else is insignificant anecdote. One has learned nothing, but one has proven something (for example that animals *anticipate* a reward, which muddies the behaviourist schema). Certainly a multiplicity of 'facts' can suddenly be forgotten when what used to be denied turns out to be that which must now be taken into account. But those forgotten facts will be replaced by facts of the same genre, responding to the same criteria of scientificity favoured by the 'serious' journals on which researchers' careers depend.

It goes without saying that meticulous and delicate thinkers will find much to criticise in the preceding descriptions. And it has to be said they are not descriptions at all, but a somewhat brutal attempt to shake up our routines, such as the idea that, aside from the ritual complaints about research being too compartmentalised, or the need for inter- or trans-disciplinarity, our research institutes, before their recent dismantling, were a first approximation of a healthy division of labour corresponding to the zealous obligation to advance knowledge. More precisely, mine is a thought experiment responding in the end to a simple enough hypothesis: namely, that the type of knowledge associated with the notion of modern science would have the very peculiar feature of not being first and foremost discursive, fitted with 'and therefores' that allow one to move from one statement to the next. Rather, this knowledge would turn every 'and therefore' into something that is only of value to the extent that it communicates with the event of a successfully created relation, which means that its value is a matter of suspense. This is what my thought experiment was trying to explore: the possibility that dynamic collectives of knowledge-construction might come together in an apprenticeship to this art of creating 'suspense'.

Obviously this experiment was not particularly programmatic, but I tried to hypothesise by offering what Whitehead calls a 'lure' for thought and the imagination. I wanted to generate the thought, and the feeling, that we have no idea what our sciences might be capable of, or could have become capable of in a slightly different world, where the value of what a scientist 'reports', as evaluated by her competent colleagues, communicates with a new kind of realism. This is the exploration of what reality asks for when what needs to be reported about it is indissociable from what it has compelled us to learn.

Symbioses

One thing is sure: this world that would be 'a little different' is not a world in which 'pure' science would be respected, as the pure effort of Man getting up on his hind legs and deciphering, one after the other, the enigmas of the universe around him. Ever since the beginnings of modern science, scientific knowledges have participated in the creation of 'non-scientific' values; and the idea of the 'temple of science' – which, according to the picture Einstein used to paint, would welcome anyone who wants to flee the mediocrity of the mundane world to discover a profound intelligibility – has communicated with the ideal of a contemplative kind of truth that has strictly nothing to do with what makes the modern sciences unique.

And yet, what we call 'valorisation' – scientific knowledge being valued for reasons other than its contribution to the 'advance of knowledge' – must certainly escape the twin models of experimental and cameral sciences, while at the same time permitting us to describe these models as particular cases. Here I shall employ the notion of symbiosis as a joining of heterogeneous beings – where each has its respective world matter in heterogeneous ways, from which each benefits, or which each valorises, in its own way.

The history of the experimental sciences offers numerous examples of symbiosis: with mathematics, with technology, but also with those who have the power to 'valorise' what they make. This also applies to the cameral sciences, whose conventions are constantly under review in response to the continual transformations in regard to what is legal, regulated, prohibited or monitored. No 'should', 'can', 'cannot' or 'must' can ever be reduced to the 'and therefores' that flow from a scientific proposition. They are always the product of negotiations, concerning definitions of prosperity or order, according to the circumstances, and between interests with more or less power.

But this also shows how symbiosis is always susceptible to being flipped into a pure and simple relation of capture. The contemporary murder of the goose who laid the golden egg – who thought her eggs were indispensable, and valuable enough to save her from the imperatives of competitive flexibility – is there to remind us. What is interesting about the notion of symbiosis is that it communicates with both a pluralisation of modes of 'valorisation' and with an active attention focused on the danger of capture.

The symbiosis between science and technical-industrial innovation has now flipped into a straightforward relation of capture. But as we have seen (and this theme will be amplified in the two following chapters), it was characterised for quite a while by a radical reduction in the number of protagonists allowed to assess the 'value' of scientific propositions. Conversely, if this value should escape capture by the order-words of progress and modernisation, then the term 'valorisation' should become synonymous with 'problem' and demand full investigation. In this perspective, the idea of 'slowing down' the sciences communicates with the question of how scientists might be trained to take part in such slowing down, notably by challenging all those modes of appreciation and judgement through which they are supposed to take on board their duty 'not to waste their time'.

Since the issue of symbiosis can hardly be stopped there, I would like to finish this chapter by fabulating another type of symbiosis, precisely where we have every reason to think in antagonistic terms. Imagine the social sciences 'disamalgamated' from the cameral sciences, affirming the highly selective character of the situations where 'learning from' is possible. This would indeed include the necessity that the addressee (from whom the investigator is supposed to be learning) be empowered to evaluate the way they are being addressed, and to do so without trying to 'capture' the investigator in the process, making her into their spokesperson. This double condition is a symbiotic interlinking. Both the 'visiting' investigator and her hosts should be capable of agreeing not to capture each other. If this condition is met, then they are likely to learn things that matter to them, but in different ways. What such social sciences require is also what democracy, as we call it, requires, if it is identified with a collective dynamic that allows those concerned with an issue the capacity not to accept or to defend a ready-made formulation of it. So the social sciences would be in a symbiotic relationship with processes through which groups become capable of formulating their own problems. And it is here that one is tempted to think in antagonistic terms about our relation to State reason, or what is today called the practice of (good) governance. I would like to try to think about this antagonism in non-essentialist terms, as resulting from an operation of capture, which thus also implies that there is, or was, a possibility of symbiosis.

Take, for example, the issue of the research evaluation process, or more generally the evaluation of any meaningful practice where the addresses are likely to contest, if they are empowered to do so, the relevance of the questions asked of them. It may be a legitimate governance decision in the general interest to make an evaluation necessary (for example, prescribing the necessity of clinical testing in toxicology). In the case of research, it is also legitimate to acknowledge that, as we have seen, the kind of evaluation performed by competent colleagues has become ineffective. The new public management discourse, however, proposes an answer to evaluation problems that expresses the way governance itself has been captured: it has been redefined in terms of competitiveness and flexibility (in the name of growth). If they had not been so captured, governance and the cameral sciences would certainly not have been able to frame the issue in terms of the relevance of evaluation, because they are not in the business of relevance. Left to themselves, they would perceive any situation according to their own categories: 'should be subject to evaluation'. The possibility of a non-defensive response – no evaluation! – requires the negotiation of ways of evaluation, and such negotiations require 'recalcitrance', i.e. the capacity of those concerned to formulate what matters to them, what an evaluation should take into account, and what would constitute an acceptable 'convention'.

Let me be quite clear. When we ask the question, 'How do we want to be evaluated?', it is a real test requiring the collective dynamic of empowerment that I associated above with democracy.[9] And it is obviously here that the social sciences could both learn and valorise their knowledge in an environment where that knowledge would not be an authority but a resource – not 'against' governance, but in a way that activates possibilities for resisting cameral capture. The link between the social sciences and the State would be neither antagonistic nor collaborative, just a link as precarious as the very definition of a 'democratic State'. It would unify two ways of making things matter, each being the other's nightmare in its own way. The social sciences should never dream of being the State's best friend; the successes of these sciences are more likely to make life complicated for it. But the way in which the State anticipates and expects such complication, or indeed suffers and tolerates it at best, is a measure of the effectiveness of its relation to what we call democracy.

Elinor Ostrom's work is a contribution to this type of social science. Ostrom complicates the supposedly incontrovertible idea that a resource likely to be overexploited by its users should be protected either by public regulation or by privatisation (in his own interest, the owner is supposed to take care of it. . .). She shows how this idea presupposes that users are defined in terms of an aggregate of so-called individual behaviour. Each individual, even if they have personal scruples about overexploitation, will refuse to be the 'altruistic victim' while others are taking advantage and profiting egotistically from the resource. Ostrom studied the way in which the behaviour of groups, in many places, contradicts this presupposition, as well as the way in which the capacity of other groups to do so has been destroyed by the 'well-meaning' intervention of the powers that be. On the basis of these empirical enquiries, she defines the conditions that make possible the functioning of what is generically called 'the commons'.[10]

Overexploitation certainly constitutes a general case, but its generality changes direction: it results from a process of expropriation, from the destruction of what makes a group capable of a form of collective intelligence, one consequence of which is the satisfaction of the conditions defined by Ostrom. This is a consequence and not an aim: it is important to stress that Ostrom's conditions don't account for the effective capacity of the commoners not to destroy what they depend on. They are the *sine qua non* actualisation of this capacity. In other words, Ostrom didn't have a 'better understanding' than the groups themselves of what made them capable of succeeding in not overexploiting their resources. Looking at the success common to these groups, she came away with a lesson, rather than a recipe, a lesson addressed to those who have the power to destroy this capacity for success.

It is an important distinction, because we are used to the extraction and implantation operations by way of which the experimental sciences identify what earlier techniques did 'without knowing it'. Such operations thus make 'modernisation' possible, a re-implantation in a new milieu, not a 'purified' one. Earlier meanings will have been eliminated, but new meanings (profitability, competition, etc.) will be introduced. This kind of operation, however, demands successful extraction, not the self-appointed right to separate what one judges as important from what is defined as illusion. For example, when cognitivists define the notion of competence as what is 'really' important, whatever the 'illusions' of teachers, and pedagogues appropriate the notion to apply it to a school context, they are convinced they are 'modernising' pedagogy, and therefore bound to make it more efficient and democratic. The least one can say is that this operation doesn't work, and that the same would probably happen if, with misplaced good will, the conditions extracted by Ostrom were to inform similar 'application' projects, thereby short-circuiting the question of what makes a group hang together, how it makes its world matter, or how the beings that inhabit this world matter for it.

Here again, the model of symbiosis between the research laboratory and the 'development of productive forces' is a bad one. This does not mean at all that the idea of extraction should be proscribed in itself. The sciences function through extraction, through a process of apprenticeship in which something implanted in one place is extracted and reported to others for whom it will make sense. What poses the problem is the way in which extraction and modernisation have been linked, transforming the question, 'What can we learn here?', into a principle of judgement that identifies what has been extracted with what really matters, and relegates the rest to an overlay of beliefs and parasitical habits. A genuine prohibition is needed to dissolve this link: *no one should be authorised to define generally 'what really matters'*. This is not a moral prohibition, but a condition of symbiotic culture, of a culture in which the capacity of each protagonist to present what matters *for them* is important, and where each will know that what they may learn from the other will always be understood as a response to a question that matters for them. Our questions are ours. Their value relates to relevance, of course – which demands that they are not unilaterally imposed, and that the response is not extorted – but it is precisely relevance that banishes the dream of extracting what is 'really important' despite what the other may happen to 'believe'. One doesn't disqualify something one depends on. If what makes the other exist in their proper consistency is what permits their recalcitrance, and if recalcitrance is a condition for the apprenticeship towards relevance, then the dream in question relates not to the adventure of the modern sciences, but to the happy times of colonialism, when the people were, along with everything else, resources from which we had to extract whatever would allow us to 'progress', and, as it happens, to say 'they believe, but we know'.

Slowing down . . .

The right to 'slowness' is not an end in itself, nor does it refer to the way some researchers ask to be 'left alone' so they can continue to think of themselves as entitled to privileged treatment. Rather, in the sense I have outlined here, slowness, like speed, has a meaning which links researchers to all those who know that the imperatives of flexibility and competitiveness condemn them to destruction.

The stakes inherent in such destruction may evoke the period of the enclosures, when peasant communities were not only robbed of vital resources, but also separated from what held them together. With the commons privatised, what was destroyed was practical know-how, along with collective ways of acting, thinking, feeling and living. If capitalism today seems to be getting along very well with modern States, it is because both are rooted in this kind of destruction. The democratic individual, the one who says, 'It's my right . . .', is the one who takes great pride in an 'autonomy' which, in fact, hands back to the State the responsibility for 'thinking through' the consequences. A strange liberty it is not to have to think further than one's own immediate interests. As for capitalism, it is running free in a world exposed to its redefinitions, all of which intensify our dependency on modes of production that presuppose and entail, as with the enclosures, a form of 'progress' that destroys all possibility of collective intelligence – as research institutes, in the wake of so many others, are discovering today.

To speak of destruction is to speak of a resistance that can only exist alongside what American activists call 'reclaiming' – recuperating, healing, becoming capable once again of linking with what we have been separated from. This 'recuperation' process always begins with the jolting realisation that we are well and truly sick, and have been for a long time, so long that we no longer recognise what we are lacking, and think of our sickness, and whatever sustains it, as 'normal'. What I have tried to do, in relation to the particular case of scientific research and evaluation, is to start thinking about what is lacking, about the way this lack makes us sick. We may well be critical and lucid, but we are crucially incapable of resisting what is destroying us (like those users who are incapable, as individuals, of not abusing a common resource).

Knowing that one is sick creates a sense of the possible. We don't know what the strange adventure of the modern sciences could have been, or could yet be, but we know that doing 'better' what we are already in the habit of doing will not be sufficient for learning. It is a matter of unlearning an attitude of more or less cynical ('realist') resignation, and becoming sensitive once again to what we perhaps know, but only as in a dream. It is here that the word 'slow', as used in the slow movements, is adequate. Speed demands and creates an insensitivity to everything that might slow things down: the frictions, the rubbing, the hesitations that make us feel we are not alone in the world. Slowing down means becoming capable of learning again, becoming acquainted with things again, reweaving the bounds of interdependency. It means thinking and imagining, and in the process creating relationships with others that are not those of capture. It means, therefore, creating among us and with others the kind of relation that works for sick people, people who need each other in order to learn – with others, from others, thanks to others – what a life worth living demands, and the knowledges that are worth being cultivated.

4

Ludwik Fleck, Thomas Kuhn and the Challenge of Slowing Down the Sciences

It is, of course, quite traditional to compare Ludwik Fleck to his 'discoverer', Thomas Kuhn. My approach will be a bit less traditional in that I will not treat this contrast as a question belonging to epistemology or the history of thought, but rather as a test, a bit like a chemist who test her compounds using different reagents. The reagent I will add is the connection, which both Fleck and Kuhn proposed, between the question, 'What is a fact?', and the question of what collectively matters for the particular community for which it is a fact. In other words, in my approach, neither epistemology nor philosophy has the exclusive right to define 'facts', nor does sociology have the right to equate facts with some social convention. The answer to the question, 'Is it a fact?', belongs to those for whom this question is a matter of concern.

Twenty years ago, the idea of a 'social construction of facts', as taken up by critical thinkers, became associated with 'relativism' by the scientists it infuriated.

However, as we shall see, the way the so-called 'knowledge economy' is mobilising research today may be equated with the possibility of a victory for relativism. That mobilisation is proceeding through the destruction of the collective, cooperative dynamics associated with scientific progress, dynamics which Ludwik Fleck first described in his *Genesis and Development of a Scientific Fact*,[1] a description which inspired Thomas Kuhn's famous *Structure of Scientific Revolutions*.[2] Putting both Fleck and Kuhn to the test of this new configuration, I will propose that scientific thought collectives, facing the prospect of their destruction, should actively accept that their concern for 'facts' must include the way these facts come to matter for other collectives.

I will begin my exploration with Thomas Kuhn because in my own life I have had first-hand experience of the kind of thought collective he characterised. Indeed, I first read Kuhn just after I got my Master's degree in chemistry, when I turned to philosophy and began exploring the resources of my new field. I thus read *The Structure of Scientific Revolutions* fresh from the experience of receiving a scientific education, and thought that here at last was a realistic rendering of the way students become part of the kind of disciplinary community characterised by Kuhn as 'normal science' – a community working within a paradigm that it does not feel the need to question, or even see the possibility of doing so.

In fact, I turned to philosophy precisely because I felt unable to comply with the strict division between the productive, scientific questions and the 'idle' ones, or those that concern the philosophers. I experienced the invisible normativity that Kuhn associates with paradigms, the common sense role they play, the thoughtless sharing of what it is that defines one's community belonging.

I believe my reading was fairly typical of physicists and chemists. They accepted and endorsed the notion of a paradigm as a way of understanding the cumulative nature of the progress made in those scientific fields where practitioners agree on the criteria for the kind of questions to be asked of what one addresses, for the tools used in asking those questions, and for what will constitute an acceptable answer.

However, as we know, the reception of Kuhn's work has been a complicated process. In order to characterise its reception by different thought collectives, it is helpful to employ the notion of a 'matter of concern', as a case of what Ludwik Fleck would call 'intercollective interactions'.

The first collective to react was, of course, that of philosophers of science, who were scandalised by Kuhn's claim that paradigms were incommensurable. The idea that there are no neutral facts to support a comparison between rival paradigms was perceived as an offence against their own self-appointed role as the keepers of scientific rationality, as those whose task it is to extract, and think with, the rational norms that must be respected in order to ensure the progress of scientific knowledge. This reaction is testimony to the difference among thought collectives. For these philosophers the fact that scientific communities were able to decide for one paradigm against another, without reasons they themselves would be able to recognise as rational, was equivalent to interpreting such a decision in terms of mob psychology.

In the 1980s, however, new thought collectives entered the scene. Each had a distinct agenda, but they shared a common concern. Whether they were critical theorists, feminists, specialists in post-colonial studies or in a new brand of sociology of science, they considered it crucial for their agenda to show that the sciences were a social practice like any other. Agreement among scientists was just an agreement among social protagonists about a reality that was unable to make any difference whatsoever; unable, that is, to be attributed any responsibility for such an agreement. It was above all the paradigmatic sciences that were to be targeted by this claim, since other fields showed all too clearly their dependence on human ideas and methodological choices.

So, Kuhn's paradigms became the royal road towards an inclusive, relativist understanding of the sciences, demoting their universalist claims. If even a science like physics had no privileged access to a reality able to force agreement onto every rational person, then it followed that all knowledge must be a social construction. The way was thus free for each collective's agenda: for the struggle against the imperialist disqualification of non-modern ways of understanding nature; for the quest for feminist epistemologies; or for sociological explanations of the triumph of one paradigm over another.

As we know, the scientists' response to the implications of Kuhn's work was in marked contrast. They did not share the philosophical concern about incommensurability, but they felt attacked by the critical claim that their access to reality could be reduced to a mere social agreement. Their reaction has been dubbed 'the science wars', and, however rude the warriors' arguments may have been, I maintain that we should take them seriously. More precisely, we should bear in mind that the kind of thought collectives Kuhn described don't really care about the idea that they exemplify a particular process of acquiring rational knowledge. What they were really concerned about, in contrast, was the social constructionists' claim that what they address is irremediably mute, that is, unable to distinguish between different ways of understanding it. What they rejected was the claim that each thought collective has its own way of 'seeing' reality. Incommensurability was not a problem for them as long as it meant only the absence of a neutral meta-position from which the merits of two rival paradigms could be assessed. It only became a problem when incommensurability was taken to mean that all ways of knowing must be recognised as somehow equivalent.

The physicist Steven Weinberg, who was to become one of the main proponents in the attack on cultural, relativist conceptions of science, wrote at the time how surprised he was to see that Kuhn had become the foundational reference for his enemies, a role that Kuhn himself had never meant to play. Reading Kuhn again, one can only be impressed by the deep ambiguity of his text, a real duck-rabbit text, to refer to the famous optical illusion (see figure overleaf).

Scientists like myself, who endorsed Kuhn's description, at first saw the rabbit of the radical distinction between paradigmatic and non- or pre-paradigmatic sciences, which explained for them how their own sciences benefited cumulative progress, while other modern sciences, try as they might to mimic this, could not achieve it. But then they discovered the relativist duck empowering their enemies.

Source: Fliegende Blätter, 23 October 1892/Wikimedia Commons

The rabbit reading is not troubled by seeing the problems dealt with by 'normal' cumulative science as puzzles, and the solutions as conforming to the paradigm. Nor is it troubled by the incommensurability among paradigms. Researchers know all too well that the road from perceiving to successfully solving what Kuhn called a puzzle – when what the paradigm anticipates is actually verified – is hard and demanding, full of colleagues ready to object to any shortcut that would evade or blur the issue.

For these rabbit readers the very existence of insistent and resilient anomalies that often play a crucial role at the outset of a scientific revolution is proof enough that Kuhn was not placing interpretation at a vantage point where it has unilateral control. There would be no anomaly if experimentation could force interpretation onto a mute or confused situation. Scientists would feel free to adopt some *ad hoc* mutually agreeable solution to explain away the difficulty. For these readers, what Kuhn had shown was that objections that put the interpretation of a fact to the test are relative to an epoch. But what matters for them is not the idea of the atemporal authority of a paradigm, but its capacity to guide the production of facts which have the power to make competent, ready-to-object colleagues agree on its interpretation. These colleagues will share the same paradigm, certainly, but they will also demand a solution from each puzzle that effectively verifies the authority of the paradigm, that is, that demonstrates that this authority has not been arbitrarily imposed on a situation which was not in fact able to sustain and confirm it.

As for incommensurability – or why it might be impossible for scientists to agree on which test will make an authoritative difference between competing paradigms – it did not mean at all, for the rabbit readers, that such a difference cannot be *created*. And Kuhn indeed characterises the period following the proposition of a new paradigm as being dominated by a collective process of critical discrimination, during which scientists work upon the creation of such a difference. This process entails the active production, exploration and evaluation of the diverging consequences of the two paradigms through the invention of experimental situations that will mark the difference between their respective performances and so allow the evaluation of their respective fecundity.

This is why for Kuhn, as for his rabbit readers, the tale was never one of arbitrariness or 'mob psychology', but one of a competent and passionate hesitation in a matter of crucial concern, a matter on which researchers were prepared to bet their reputation, their future work, and the future of their field.

But then came the duck reading, which took advantage of certain other aspects of Kuhn's account.

If Kuhn's first concern was to resist an ahistorical definition of facts, he indisputably downplayed the exceptional character of the achievement of the kind of facts which inspire the trust of puzzle-solving scientists, those facts whose interpretation will resist the objections of competent colleagues. When the possibility of an anti-realist reading of Kuhn was developed, his horrified rabbit readers discovered that nothing in his text explicitly opposed reducing this achievement to a mere social agreement. Worse, Kuhn's explicit extension of his notion of the paradigm to fields like Aristotelian physics or pre-Copernican astronomy completely contradicted his very sharp discrimination between contemporary paradigmatic and non-paradigmatic sciences.

It may well be that in both cases Ludwik Fleck's influence on Kuhn played a role. Indeed, Fleck's characterisation of the scientific thought style as aiming at minimising 'thought caprice' while maximising 'thought constraint' under given conditions is relevant for *any* scientific collective, including those that don't work under a paradigm. Fleck's idea of a fact – such as the factual relation, in one of his great case studies, between syphilis and the Wasserman reaction – as being what puts a stop to free arbitrary thinking, seems to correlate well with the authority of paradigmatic cases. This stop, Fleck writes, 'must be brought home to each member as both a thought constraint and a form to be directly perceived'.[3] The duck reading of Kuhn would then be made possible by his acceptance of such a characterisation, by his not taking as an active thought constraint the differences between facts in sociology, biomedical research, pre-Copernican astronomy or quantum mechanics. In so doing, Kuhn blurred the dramatic distinction he himself had proposed: such facts may well look similar, but what Fleck and I would call their natural history is not. Correlatively, the communities that bring them into existence, and which they organise, also offer interesting contrasts.

Supporting this hypothesis is Kuhn's recognition that he had no answer to the question of why only some sciences become paradigmatic, while others do not, even if they try to acquire a paradigm – and we know that, after Kuhn published his study, many tried desperately to do so, but without much success. It seems to me that the very fact that Kuhn asked this question indicates how much he had relied on Fleck for the characterisation of a fact as what resists caprice, or arbitrary free thinking. Why, then, wouldn't all scientific collectives equally benefit from this thought-style?

I would thus claim that rabbit readers like myself automatically added to Kuhn's text the exceptional character of the achievement underscored by the authority of a paradigm. This authority expresses itself as an 'event' which succeeds in actually grasping reality. The scope and meaning of this grasp may well change as a consequence of a scientific revolution, but will not itself be dissolved. Isn't one of the constraints on a new paradigm that it must explain and reassure the older paradigm about the reliability of its experimental equipment? A paradigm does not disappear like a dream, but lingers on in laboratory instruments. What those instruments have reliably established will be different but still significant.

In fact, I would claim that scientists working with a Kuhnian paradigm would never admit that the thought constraint a fact achieves is of the nature of a stop. More precisely, under the conditions defined by the paradigm, what established facts should have demonstrated their power to 'put a stop' to is not free, capricious thinking but *objections*. They have to earn recognition as what I call 'reliable witnesses', witnesses that authorise one way of understanding over possible other ways. Correlatively, the acceptance of such facts is a matter of intense collective concern. Indeed, as reliable witnesses, they will act as a dynamic constraint for the collective, opening up the possibility of new questions, new experimental settings, and new puzzles.

From this point of view, the reliable detection of syphilis appears only as an empirical success. The reliability of the Wasserman reaction studied by Fleck is certainly important for medical reasons. But the fact that syphilis is likely to be detected by this test does not authorise a particular interpretation of the disease. It does not impose constraints on Fleck's researchers that would lead to a cumulative process of acquiring knowledge about it. It is indeed just a stop, not a collective 'go'.

I have just introduced the concern which situates me. As I was educated in chemistry, and as such shared the rabbit reading of Kuhn, I came to be impressed by the radical plurality of thought collectives unified under the category of 'modern science'. This category unifies a spectrum of practices the two extremes of which, for me, have nothing in common. At one extreme, we find researchers who are empowered by belonging to a collective, researchers who passionately imagine, object to and test hypotheses in constant interaction with those colleagues on whose interest and objections they depend. At the other extreme, we find fields in which the prime matter of concern is the conventional imposition of the accepted methodological strictures of objectivity on whatever is addressed. No puzzle here, and no possible anomaly; rather, the sometimes difficult elimination of anything that endangers the scientific character of the facts obtained.

We are thus dealing with two very different ways of minimising what Fleck called 'thought caprice',[4] corresponding to two very different collective dynamics. In the first case, objectivity is defined as a collective achievement that requires cooperation, implying that objections are a positive, even necessary and called for, part of the collective game. In the second, each individual work attracts a rather suspicious, censorial mode of attention. Here collective interaction is about the correct application of the method, with no special interest in the facts themselves, each being added like a brick to an edifice rather than assessed in terms of the new possibilities or questions it allows us to envisage.

And in between these two extremes of the spectrum are fields that resemble Ludwik Fleck's own, struggling with questions of public interest, challenging and challenged by what I would call the messiness of the world.

Re-reading Fleck I was touched by his beautiful characterisation of the precarious character of the grasp biomedical researchers have on what they address. No paradigm here, because there is no puzzle. I loved Fleck's gentle humour addressed to the rigid thought-style associated with Pasteur and Koch. Both tried to institute what Kuhn would call a paradigm, but were unable to do so because each disease, each microorganism, each culture, never stopped introducing unpredictable questions of its own, demanding clear-headed attention rather than the confidence of the puzzle-solver. I believe that just as Kuhn could be appreciated by physicists and chemists, today's biotechnologists and biomedical researchers would understand, and maybe secretly share, Fleck's humour, even if they feel bound to think of their science as conforming to the grand model of cumulative paradigmatic science.

I would say that, while Kuhn's paradigm was organised around the question of the cumulative dimension of some sciences, Fleck's questions address fields in which facts usually cannot be given the power to authorise a single interpretation because of the intrinsic and entangled variability of what is addressed in those fields. When Fleck writes that the researcher 'gropes but everything recedes, and nowhere is there a firm support',[5] he is not making a general epistemological point. He is giving voice to a pragmatic assessment of the 'reality' at stake in his field, a reality that will disappoint those who believe in the authority of facts. And when he asks why 'all rivers finally reach the sea, in spite of perhaps initially flowing in a wrong direction, taking roundabout ways, and generally meandering',[6] he is asking a real question. We know his answer: the rivers do not reach the sea as if the sea had something special about it. The lines of research do not 'find' an answer they will agree upon. 'Provided enough water flows in the rivers and a field of gravity exists, all rivers must finally end up at the sea.'[7]

The retroactively described cumulative development that finally led to the Wassermann test needed such flowing water, that is, the continuous cooperation and mutual interactions of the members of a collective. But this water would have remained dispersed in a thousand rivulets if syphilis had not been a matter of public concern, if there had not been an 'insistent clamour of public opinion for a blood test'.[8] The clamour raised by the syphilis epidemic was the field of gravity that was needed, providing the dominant and directing orientation necessary for old and new lines of thought to develop, join up, be modified by each other, merge and finally produce what would be retroactively acknowledged as a 'real finding'.

This contrasts strongly with Kuhn's emphasis on the need for the autonomy of research questions with regard to their social value or interest. For Kuhn, the paradigm is what determines the right questions. What comes from the outside and cannot be identified as a puzzle would only perturb the course of normal, cumulative, science. So paradigmatic sciences have to be protected from the expectations they may kindle in society. More precisely, they have to both kindle them and disallow them.

A parallel contrast characterises the situation where a researcher is consciously aware of the particular thought-style he shares with his collective. For Fleck, the difficult and always partial character of such awareness is an empirical fact, while for Kuhn it is not. For him, researchers *should not* be aware of the constraining power of their paradigm, or else they will lose the puzzle-solver's tenacious confidence. Lucidity, then, is the enemy of scientific creativity.

Let us now turn to the present-day knowledge economy. At this conjuncture, not only is the autonomy of the research community, which Kuhn considered crucial, coming to an end, but this also throws into question the Fleckian distinction between *esoteric circles* – of specialists who are 'in the know' – and *exoteric circles*, whose members share and support their thought-style with 'vivid certainty' but are not empowered to actively participate in the assessment of the corresponding research. Given the number of researchers employed in industry since the nineteenth century, the distinction was always a precarious one, but the partnership now required between public research and private interests

has exploded it. Private partners can hardly be characterised as an 'exoteric' circle; they enter forcefully into
the thought collective's esoteric knowledge.

The question of the 'paradigmatic' sciences' relation
with industry is not, however, a new one. Kuhn's dramatic distinction between paradigmatic, cumulative
sciences and non-paradigmatic ones in fact replays the
matter of concern that came into the lives of chemists and
physicists during the second half of the nineteenth century. Fearing that their science would be put at the direct
service of industrial development, to the launch of which
they had so powerfully contributed, they laid claim to an
institutional landscape that would not only make possible, but indeed support, the strong division between
what Kuhn calls puzzles and all those other questions
which, however interesting, have the potential to trouble researchers, attracting them onto meandering paths
where they would no longer be guided by their paradigm.
A troubled researcher is an unproductive one. Interfering
with the fast, cumulative dynamics of the paradigmatic
sciences would kill the goose that laid the golden egg.

Today, the murder of the goose means that whatever
the differences among scientific fields, paradigmatic or
not, the new institutional landscape called the 'knowledge economy' has erased them. Only one criterion now
differentiates them, their 'attractivity', the way they fit
into the race for competitiveness and profit. And the
intensity of the deleterious effects marking the dissolu-
tion of collective research dynamics is a product of the
same criterion. Both biotechnology and Fleck's own field
of biomedicine are confronted with an explosive growth
in fraudulent or unreliable claims and largely unavoid-
able cases of conflicts of interest. The direct involvement

of the pharmaceutical industry has also profoundly transformed the relation between esoteric and exoteric circles. As Fleck himself stressed, public concern, even public outrage, was an active ingredient in the production and stabilisation of 'facts' such as that supplied by the Wasserman test. But in the new situation, the 'field of gravity' is no longer that of a public noisily addressing trusted professionals. It is provided by the pressure of multiple industrial strategies that reconfigure both the public, by segmenting it into potential profitable markets, and researchers, who are bound by patents and industrial secrecy. Disease-mongering and other market strategies are constantly creating new demands and new kinds of expectations. As for the general public itself, its confidence, as we know, is already rather deeply shaken, in particular by troubling news stories about the unforeseen effects associated with prioritised drugs that are meant to be taken not simply while one has a disease but right up until the death of the consumer.

We face a future where claimed 'facts' will accumulate at full speed but nobody will really know what is meant by a 'fact' any longer, be it a Fleckian or a Kuhnian one.

Whatever the situation in each research field, it is not surprising that a resistance movement is beginning to emerge. In 2010, a text entitled The Slow Science Manifesto was published in Berlin, which ends with these lines:

> Slow science was pretty much the only science conceivable for hundreds of years; today, we argue, it deserves revival and needs protection. Society should give scientists the time they need, but more importantly, scientists must take their time.

We do need time to think. We do need time to digest. We do need time to misunderstand each other, especially when fostering lost dialogue between humanities and natural sciences. We cannot continuously tell you what our science means; what it will be good for; because we simply don't know yet. Science needs time.
– *Bear with us, while we think.*[9]

Now, this is a fairly consensual text, which both Thomas Kuhn and Ludwik Fleck would probably have agreed with. It certainly reflects what has now become an urgent matter of concern for all scientific thought collectives. But it does not answer the concerns of those who question the kind of development that so many scientists associate with progress. It is quite significant that the authors of the Manifesto are addressing 'society' without naming who it is that is putting pressure on them, who they need to be protected from. Then there is the allusion to the hundreds of years during which scientists were given the time they needed. What we are in fact hearing here is the lament of the golden goose missing the Golden Age when scientists benefited from both autonomy and the respect due to their role in serving the general interest.

My whole point is to associate the idea of slow science with a more ambitious agenda, one that takes into account the need for a deep break with the ideal of academic science shaped during the nineteenth century, a model of research that promoted as a general ideal the fast, cumulative advance of disciplinary knowledge along with a correlative disregard for any question that would slow this advance down.

Usually the critique of disciplinary knowledge leads to

a plea for some general, interdisciplinary or even holistic thought-style. This is not my position. The opposite of 'disregard' is not 'actively including' but 'taking seriously', or 'paying attention'. Taking seriously or paying attention puts into question the way scientific disciplines have been shaped by their exclusive, quasi-symbiotic relationship with industry. And here the distinction between Kuhn and Fleck becomes crucial. For Fleck, as I have mentioned, it is difficult to pay attention to, or be aware of, the particularity of one's own thought-style and the way it selects and discards aspects of a situation which 'do not really matter', but the difficulty is an empirical fact only. For Kuhn, ignoring this particularity is crucial for the tenacious creativity of the puzzle-solver. In other words, for Kuhn, the training of gooselike researchers with their imaginations strictly and normatively channelled, as initiated by the chemist Liebig,[10] is a crucial factor in the creation of the kind of institution able to protect cumulative and inventive scientific progress from sterile meandering. If he is right, the only meaning for slow science, as a perspective of resistance to the knowledge economy, is that of 'back to the Golden Age'.

However, the esoteric-exoteric contrast, as Fleck characterised it, must also be questioned. The symbiotic relation between academic science and the industrial world does not conform to this contrast. There is nothing exoteric in the constraints and concerns of industrial production and marketing. Exoteric knowledge is the province of the general public, ensuring the 'stark certainty' of scientific results as well as the image of a science that will finally answer questions of common concern in a rational and reliable way.

This is the lie, or the bluff, that I argue has to be

called out as soon as we recognise the non-sustainable character of our development. Consequences that were disregarded in both scientific and industrial environments, far from being fixed, are putting our very future into question. Rationality and objectivity, as promoted by exoteric knowledge, have been instrumental in silencing voices coming from other thought collectives protesting about what has not taken into account by so-called rational progress. They have also been used to justify scientists' poor imagination and cultivated disinterest in the messy complications of this world, the only world we have. From the prospect of climate disorder to pollution, the poisoning of living beings by dangerous cocktails of new chemicals, and other ecological disasters, it may be said that the messiness of the world is now returning with such a vengeance that the motto 'progress will repair the collateral damage that progress has occasioned' has lost all credibility.

But slow science is not about scientists taking full account of the messy complications of the world. It is about them facing up to the challenge of developing a collective awareness of the particularity and selective character of their own thought-style. This, however, should not be confused with a call for lucid reflexivity to be developed inside thought collectives. It is rather a matter of collective learning through the test of an encounter with dissenting voices around issues of common interest. Such a learning process demands of modern collectives what I would characterise as a 'becoming-civilised'. So slowing down the sciences means civilising scientists, civilisation being equated here with the ability of members of a particular collective to present themselves in a non-insulting way to members of other collectives, that is, in a way that enables a process of relation-making.

In order to relate rather than insult, a presentation should never involve the claimed possession of an attribute that defines the other as lacking it. For instance, when a scientist defines her practice as objective or rational, she is insulting to the extent that she implies that this is a distinctive characteristic that the one she is addressing lacks. Likewise, Fleck is on dangerous ground when he characterises science as aiming at the minimisation of thought caprice; he needs immediately to add that caprice is not a general judgement but may well refer to aspects of a situation that matter a great deal to other collectives.

Presenting oneself in a civilised manner means presenting oneself in terms of one's specific matter of concern, that is, admitting that others also have their matters of concern, their own ways of having their world matter. Civilised scientists would make it public, a matter of exoteric knowledge, that the reliability of their results is related to matters of concern as well as to competent knowledge; and that the very particular conditions required by the latter come at the price of ignoring what may be important factors outside the laboratory. They would acknowledge that when what they have achieved leaves its native environment – the network of research laboratories – and intervenes in different social and natural environments, it may well be leaving behind its specific reliability. And they would recognise that restoring reliability means weaving new relations proper to each new environment, which entails welcoming new objections – no longer just the objections of colleagues, but those of other collectives concerned by aspects of the environment that the scientists themselves were not concerned with.

In other words, civilised scientists, true to the specificity of their practice, will insist that reliability is not a stable attribute, and that the 'valorisation' of a possibility born inside a research environment requires a radical redistribution of expertise through the creation of demanding new relations that will give voice to the often messy web of hard questions that matter in any given situation.

Such a redistribution cannot be thought of in terms of the contrast between exoteric and esoteric knowledge. Rather, it demands that the situation be understood through the diverse matters of concern that connect with it, with no *a priori* differentiation between what really matters and what doesn't. Such an understanding requires a kind of imagination that research collectives have not cultivated. Instead, they have systematically downplayed anything that doesn't directly contribute to the cause of advancing specialised knowledge, and have proscribed, as a waste of time, interests and questions that would enable specialists to take seriously the matters of concern arising from the innovations they promote.

This is why talk of slow science is a direct challenge to the motto structuring research collectives: *do not waste your time with idle questions, questions that cannot be reduced to scientific terms; this would be betraying your sole duty, the advancement of knowledge!* This motto promoting and mobilising fast science is the very recipe for channelling attention and eagerness, and for restraining imagination. It enforces the idea that the rational approach to situations should extract those dimensions that can be defined as scientific or objective, and leave the remainder to be addressed by other means which are not the scientists' concern . . . and, it will be added, which should *not* be their concern because they trespass on matters to be decided in terms of political or ethical values. 'Society will decide', they say, never wondering about how and by what means such decisions are taken.

Civilised scientists are not, however, scientists with a general culture. What they have to cultivate is the capacity to participate in the collective assessment of the consequences of an innovation, rather than a decision based on values. Indeed reliability 'out there' will depend on facts other than the scientific kind, brought by other, non-scientific, collectives. They may also come from objections that might be very different from those of competent colleagues who all share the same values and work in similar environments. General culture is no great help when interacting with protagonists who are not academically trained but are nevertheless empowered to object; nor is interdisciplinary culture as it has developed among certain polite academics. Where a minimum of trust prevails, even in the best of cases the process will be, and must be, slow, difficult, rich in friction, and torn between diverging priorities. Any nostalgia for clean, competent collectives made up of 'dear colleagues' will result in the conclusion that outsiders are unable to participate, that they are not partners, just annoying troublemakers.

Slow science thus represents not only a challenge to fast, mobilised science. It is also a wager. A wager on the capacity of scientific thought collectives to enter into new symbiotic relations with other collectives that have different matters of concern. The very term 'slow' is indicative of this wager. Slow, today, designates all those social movements that endeavour to escape what has been put forward in the name of efficiency, and discover that in this name many relations have been cut or destroyed, to be replaced by divisions and oppositions between contradictory interests. Slow food movements, for instance, are discovering that the interests of producers and consumers need not be opposed. Thinking together and negotiating can not only open up new, mutually agreeable transactions, but might also become important and rewarding in themselves. People come to realise that by adopting certain patterns of consumption they can help the kind of producers they have learned to appreciate. The latter, in turn, can become acquainted with those for whom they produce. Such experiences give new meanings to food.

My own wager is that fast, mobilised science is not rewarding. What is rewarding is what Fleck emphasised: that special kind of dynamic interaction that produces and activates a collective. This is why it is important to claim that slow science is not against specialised science – against scientists assembled by common matters of concern. Slow science, as I defend it, is rather expressing the trust that this specialised dynamic does not need mutilated, channelled, mobilised minds. It also trusts that scientists may well find it rewarding to participate in other dynamics and learn from their encounters with empowered collectives.

When Fleck wrote, in 1929, about natural science 'as the art of shaping a democratic reality and being directed by it – thus being reshaped by it',[11] he was probably thinking of a 'natural' reality understood as free of any transcendent authority. But in the same text he described the 'democratic way of thinking' as having first developed 'among the artisans, the seamen, the barber-surgeons, the leather workers and saddlers, the gardeners and probably also children playing . . . Wherever serious or playful work was done by many, where common or opposite interests met repeatedly, this uniquely democratic way of thinking was indispensable.'[12] It is sufficient to substitute 'the art of shaping a democratic reality and being directed by it' with 'the capacity to participate in a democratic way of thinking and learn from it' to get the formula for what I call the 'civilised sciences'.

5

'Another Science is Possible!' A Plea for Slow Science

Some years ago, many academic dissertations were written on the rights of future generations in relation to the unsustainable character of what we call development. But we now realise that the future is coming towards us at full speed. It may be said that we who are here are in the position of having to imagine how we will answer those who are not here, but who nevertheless already exist. What will we say to the children born in this century when they ask: 'You knew all you had to know; what did you do?' Any adult today might imagine being asked this question. However, as academics, I would claim that we stand in a special position.

It may indeed happen that some people, outside academia, are confident that we who are selected, trained and paid to think, imagine, envisage and propose are indeed doing so in relation to the future we face. And there may also be young people entering university in the strange hope of getting a better understanding of the threatening world we live in.

Can we consent to this trust and allow it the power to affect us? Or will we answer with the sad tale that we are, or were, really too busy meeting the relentless demands to which we now have to conform in order to survive?

I am not speaking here only of the knowledge economy and the imperative to produce knowledge that is of interest to the competitive war-games of the corporate world. Even those academic fields that don't produce patents have now been submitted to the general imperative of benchmark evaluation. They have to accept the judgement of an academic pseudo-market ruled by blind competition.

In short, we must admit that we have been successfully compelled to surrender a great part of our freedom to engage in dissent. We now have to tell our students to choose subjects that will lead to fast publication in high-ranking journals specialising in professionally recognised issues – issues which, in general, are of interest to nobody except other fast-publishing colleagues. We have to tell them that, if they want to survive, they have to learn to conform to the blinkered normative frames imposed by such publications.

So my first point is: whatever the future, research institutions are not equipped to formulate it, or even envisage it, in a way which would meet the trust some people may still be naive enough to place in us.

But we also know that everywhere the same disempowering processes are at work. Everywhere a similar cut-off is introduced, separating people and collectives from their capacity to envisage, to feel, think or imagine. Everywhere the same kind of attack has been launched, which can be characterised as a form of sorcery that obstinately, sneakily and wickedly paralyses our capacity to resist.

This is why, faced with our lack of resistance, I will not speak of guilt. I prefer to speak of shame, remembering Gilles Deleuze's remark that 'the feeling of shame is one of philosophy's most powerful motifs'.[1] Such a motif may be extended well beyond philosophy, to all of us who may feel this shame.

I would claim that the kind of future we face creates what William James called a genuine option, an option which cannot be avoided because there is no place to stand outside of the alternatives of either consenting to, or refusing, the challenge it offers.

The process of the destruction of the academy is not in itself sufficient to create such an option. Ten years ago I was ready to admit it was a dying institution, richly deserving its fate. Today, however, this destruction can be seen, along with innumerable other such destructions, as systematically eradicating resources that could address the future, and systematically cutting off our capacity to think, that is, to escape despair and cynicism. One way or another, much of what is being destroyed may be characterised, like the academy, as deserving of its fate, but the meaning of such a characterisation has changed. It has become a way of refusing the challenge we are confronted by.

I would name that challenge 'barbarism', as the most probable outcome of what is going on today.[2] We know the taste of this barbarism already, in the so-called 'difficult but sadly necessary' measures that authorities of all kinds demand we accept, with consequences that would have deemed unthinkable yesterday. Such consequences, which we already know only too well, will only multiply and intensify in the future. This is just beginning.

Accepting that one must think, feel and imagine the necessity of facing up to barbarism means refusing the idea that other, more deserving figures will arrive to turn the tables. Today, messianic perspectives are tempting, even fashionable, but waiting for salvation from some Great Outside only plays into the hands of barbarism, by evading the challenge as it is addressed to us now.

My intervention takes 'slow science' as a name for the challenge that is addressed to us as academics. A name which also includes a trap we have to resist; namely, the call for an agreement to go 'back to the past' as expressed by The Slow Science Manifesto, discussed in the previous chapter. As we saw there, it concludes by asking an unspecified audience to leave scientists alone: 'We cannot continuously tell you what our science means; what it will be good for; because we simply don't know yet. Science needs time – *Bear with us, while we think.*'

Resisting consensus always exposes us to sniggers, but I will expose my position even further by daring to defend the definition of the task of the university given by the mathematician and philosopher Alfred North Whitehead in 1935: 'The task of a university is the creation of the future, so far as rational thought, and civilised modes of appreciation, can affect the issue. The future is big with every possibility of achievement and of tragedy.'[3]

We may snigger indeed, because it is all too easy to deconstruct the very idea that universities ever had such a task. But this is precisely the meaning of William James' notion of a genuine option. As I remarked earlier, the destruction of academia is not in itself sufficient to create such an option. Those academics who just ask for time to think – who do not name those putting pressure on them, preferring to address 'society' and ask for protection – do not feel there is an option at all. They just dream of a past where they, and the so-called disinterested knowledge they produced, were respected. The 'exposing oneself to sniggers' option requires us to accept that we academics are, among many others, called upon by our role in the creation of the future. We cannot evade that call by pleading that we do not deserve to play such a role.

Moreover, what I find interesting in Whitehead's seemingly innocuous proposition is that it associates the future neither with the advance of knowledge nor with progress, but rather with radical uncertainty. We do not know what our future will be, and nor do we know if, or to what extent, what he calls rational thought and civilised modes of appreciation can affect the issue. But this is why his proposition is relevant today, more than ever.

I will first emphasise that, already in 1935, Whitehead's proposition was something like a plea. Indeed, what turned him from the mathematician he was into the philosopher he became cannot be disentangled from his deep feeling of anxiety about the effects of what he characterised as an important discovery marking the nineteenth century: 'the discovery of the method of training professionals, who specialise in particular regions of thought and thereby progressively add to the sum of knowledge within their respective limitations of subject'.[4]

Let me make it clear, right from the beginning, that the point is not to criticise specialisation or abstraction.

Whitehead was a mathematician, and for him, you just 'cannot think without abstractions'. He would never have criticised the way the sciences abstract what matters for each of them from an always-entangled world. However, for him rationality was not the capacity for abstraction, it was rather the ability to be vigilant about one's abstractions, to not be blindly led by them. As we should recall, a good craftswoman does not know only how to use her tools, and will not look at a situation in terms of the demands of the particular tool she is used to. Rather, she will judge the fitness of the tool for the situation. For Whitehead, it is the same with the exercise of thought – you need to be vigilant about your modes of abstraction.

This vigilance is precisely what is lacking among those whom Whitehead characterises as professionals, with their 'minds in a groove':

> Each profession makes progress, but it is progress in its own groove. .... The groove prevents straying across country, and the abstraction abstracts from something to which no further attention is given. .... Of course, no one is merely a mathematician, or merely a lawyer. People have lives outside their professions or their business. But the point is the restraint of serious thought within a groove. The remainder of life is treated superficially, with the imperfect categories of thought derived from one profession.[5]

As such, professionals, fixed persons with fixed duties, are not new to the world. However, Whitehead continues, 'in the past, professionals have formed unprogressive castes. The point is that professionalism has now been mated with progress. The world is now faced with a self-evolving system, which cannot stop.'[6] One cannot stop the clocks, as Pascal Lamy once remarked.

While Whitehead does not object to the professionals' specialisation, he characterises them as 'lacking balance'. Their training, while neglecting 'to strengthen habits of concrete appreciation of the individual facts in their full interplay of emergent values',[7] leaves them prey to the power of a particular set of abstractions, promoting a particular value. I rather like the 'lacking balance' formulation, for its affinity with the image of the 'sleepwalker' that accompanied the invention of the method of training scientists as professionals during the nineteenth century, at the time when what I call 'fast science' was being invented. Whitehead's plea regarding the task of universities was thus also aimed at a 'slowing down' of science, which is the necessary condition for thinking with abstractions rather than *obeying* them.

I turn now to the invention of this type of training, which has become the general model in our universities. It is strikingly illustrated by Justus von Liebig's radical redefinition of what it is to be a chemist.

In the 'chemistry' entry of the Diderot and d'Alembert *Encyclopaedia*, the chemist Gabriel François Venel had characterised chemistry as a 'madman' passion. It took a lifetime, he wrote, to acquire the practical knowledge of and ability to master the wide variety of subtle, complex and often dangerous chemical operations pertaining to the many arts or crafts of chemistry, from that of the perfumers to that of the metallurgists or the pharmacists. In Liebig's laboratory, by contrast, a student would obtain his doctoral degree after four years of intensive training. He would learn nothing, however, of these many traditional crafts and their operations. He would use only purified well-identified reactants and standardised protocols, and learn only the latest methods and instrumental techniques. Liebig was named the 'chemist breeder', due to the hundreds of students who were trained in his laboratory at Giessen between 1824 and 1851. Many went on to set up similar university laboratories, while others played a crucial role in the creation of the new chemical industry.

Liebig's invention of what we may call 'fast chemistry' entailed a cut, which divided not pure and applied chemistry, but rather the whole continent of chemical crafts on the one side, and, on the other, both academic research and the new network of industrial chemistry, the two entertaining a new symbiotic relation, as each needed and fed the other.

Symbiosis, however, is a balance that must be maintained. It is striking that Liebig, who played a very important role in the development of industrial chemistry, also became, as early as 1863, a passionate promoter of the need for pure, autonomous academic research. He is the father of what we now call the 'linear model', together with the famous 'goose that laid the golden egg' argument: it is in its own best interest that industry should keep its distance from academic research, leaving the scientific community free to determine its own questions, because only scientists can tell, at each step, which questions will be fruitful, which will lead to fast cumulative development and which will result only in some empirical gathering of facts leading nowhere. For industry to dictate its own questions would be like killing the goose and losing the eggs.

We have heard multiple variants of the same argument, as a motto for the arrangement that many scientists associate with the Golden Age, when science was recognised as a free source of novelties that would lead to industrial innovation, ultimately benefiting the whole of humanity. However, some aspects of the argument are seldom developed. The first one is the division, a true class division, between scientists who work on protected academic territory and those who, in selling their labour power to industry, are usually denied autonomy and the freedom to contribute to public knowledge. The second aspect is that the goose with the golden egg metaphor hides an important feature of the role the trained scientist now plays as a fast science professional.

The official story is that the goose lays her eggs and is happy to learn that some of them have turned golden, in industrial development terms. She hopes that this will ultimately result in benefits for humanity, but she cannot be considered responsible for any misuse. She insists that her only loyalty is, and must be, to the advancement of knowledge, and thus, as Whitehead wrote, she is entitled to treat the remainder 'superficially, with the imperfect categories of thought derived from [her] profession'. This corresponds to the 'ivory tower' image of academic science, and it is reinforced by the other current image of scientific creativity, that of the sleepwalker walking on a narrow ridge without fear or vertigo because he is blind to the danger. Asking creative scientists to be actively concerned about the consequences of their work would be the equivalent of waking the sleepwalkers, making them aware that the world is a long way from obeying their categories. Struck by doubt, they would fall from the ridge into the morass of turbid opinions. They would, that is, be lost for science.

This image of scientific creativity as, in Whitehead's terms, intrinsically lacking balance, is deeply ingrained in fast science education. One way or another, explicitly or not, scientists learn that questions which concern the wider world, the world where the golden eggs will make a difference, should be globally defined as 'non-scientific', even if such questions are the object of a lot of scientific work in other departments dealing with cultural, social or economic problems. Interest in the world we live in becomes like a temptation that researchers who have 'the right stuff' should be able to resist.

Fast science refers not so much to a question of speed but to the imperative not to slow down, not to waste time, or else. . . . It may be tempting to associate this 'or else', which evokes the prospect of a fall, with the noble demands of a vocation, which scientists would betray if they did not devote their whole life to its fulfilment. However, the way this so-called devotion is obtained and maintained, through a training that channels attention and eagerness while restraining imagination, has nothing noble about it. What Whitehead called the training of professionals rather refers to the kind of induced anaesthesia generated by a mobilised army on the move, where the imperative is to go as fast as possible. Such an army does not wander and wonder. The imperative means that the landscape it moves through will be of no interest, only the obstacles it has to move around. Those in the army who complain about the damage its advance causes (destroying crops, stealing goods, raping women . . .) certainly don't have the right stuff. Such things should not slow down the advance. Soldiers must forget their attachments to their own crops, goods and wives. Likewise scientists when they dismiss a question as 'non-scientific'.

From this point of view, biologists defending GMOs,

for instance, may feel quite justified in claiming to have found a rational solution to the problem of feeding the hungry, quietly ignoring the social and economic causes of world hunger. They just show themselves to be real scientists, ignoring everything that would slow them down or put obstacles in the way of the progress made possible by their golden eggs.

But this last example is also enough to reveal what the official story has hidden. There never was an ivory tower for the goose with the golden eggs. The valorisation of their work, the link with those capable of turning their eggs into gold, has always been part of the activity of academic scientists, even if, like Pasteur or Marie Curie, their name is associated with disinterested research. The goose is also an entrepreneurial strategist. She is on the lookout for those who might draw golden consequences from what she has laid. What characterises fast science is not isolation, but rather working in a very rarefied environment, an environment divided into allies who matter and those who, whatever their concerns and protests, have to recognise that they are the ultimate recipients of the golden benefits, and therefore should not disturb the progress of science.

Already when he made the cut between chemistry-in-the-making and chemical arts and crafts, Liebig also cut chemistry off from the social and practical concerns those arts and crafts were embedded in and responding to. The only true interlocutors for the new academic chemists, the only ones who understood their language, were now those who inhabited the industrial world, also in the making. And this still corresponds to the intellectual equipment contemporary fast science training provides to scientists. They will easily break up a situation into its supposedly objective or rational dimensions and what would simply be a matter of contingent, arbitrary complications. And the dimensions that correspond to fast science's categories are rather naturally the very ones that are relevant for industrial development, since both agree on ignoring the same type of complications. No direct mobilisation on the part of industrial interests is necessary here; only this symbiotic relation between two modes of abstraction.

But today even this is no longer sufficient for the former allies of fast science. The knowledge economy is now destroying the home in which the goose that lays the eggs was protected. The relative autonomy of scientific research, secured by Liebig and his colleagues, belongs to the past. Some may be tempted to claim that it never existed anyway, given the intimate connection between academic fast science and industry. I disagree, and would claim instead that what is in the process of being destroyed is the very 'social fabric' of scientific reliability. In the future we may well see scientists at work everywhere, producing facts at the speed our new sophisticated instruments make possible; but the way those facts will be interpreted will mostly conform to the landscape of vested interests.

As all working scientists know, if a scientific claim can be trusted as reliable, it is not because scientists are objective, but because the claim has been exposed to the demanding objections of competent colleagues concerned about its reliability. And it is this shared concern that may well be destroyed if these colleagues are mostly bound to industrial interests, that is, bound by the need to keep the promises that attract their industrial partners. The maxim that may well prevail, then, is that you don't cut off the branch on which you are sitting together with everybody else. Nobody will object too much if objections to the weakness of a particular claim lead to a general weakening of the promises of a field. Dissenting voices will then be disqualified as minority views that need not be taken into account, since they spell unnecessary trouble. What will then happen already has a name: the 'promise economy', in which what holds the protagonists together is no longer a reliable scientific egg that may turn golden for industry, but glimmering possibilities the strength of which nobody is interested in assessing any longer. In other words, in the guise of the 'knowledge economy', the speculative economy, the bubble and crash economy, has succeeded in recruiting scientific knowledge production.

This is why we can sympathise with The Slow Science Manifesto's dream of a return to the Golden Age when the autonomy of scientific research was respected. But we have to remember that while the autonomy of fast science may well have protected the reliability of scientific claims, it never ensured the reliability of a mode of development that we are now shamefully forced to recognise as having been, and still being, radically unsustainable. This is by no means an accident. The reliability of fast science's results is relative to purified, well-controlled laboratory experiments. And competent objections are competent *only with regard to such controlled environments*. Which means that scientific reliability is situated, bound, to the constraints of its production. Which also means that when the eggs leave their native environment and turn golden, they will have left behind this specific reliability and robustness. What reliability they now have is no longer an issue of scientific judgement only, but a social and political issue.

For instance, airplanes are safe enough because of the existence of a consensus about the need to avoid crashes at all cost. In contrast, the concern for the sustainability of our mode of development, which is far from new, has until recently been anything but consensual. People who objected on these grounds were not even listened to, but attacked and derided as wanting to send us back into the cave! No doubt lip service was paid to the fact that some innovations may have unwanted consequences, but, it was added, technoscientific progress is bound to find a way to fix the damage. To doubt that, is to doubt progress! And, as we know, such doubt is blasphemous.

Here we can recognise an echo of Whitehead's point about serious professional thought being stuck in a groove, while the remainder of life is treated superficially. And the response of many scientists is just as superficial when they claim that it's not their fault that sustainability was not a public concern, since they cannot be held accountable for the way 'society' decides to use what they produce. This is the typical goose answer. As usual it ignores the fact that the claimed irresponsible use of their products never prevented academic scientists from associating scientific progress with social progress; from joining in with the 'back to the cave' insults; from presenting their science as offering, at last, rational solutions to problems of general concern; or from framing objections in terms of a simple opposition between science and value – as if all those aspects of a concrete situation that they are not equipped to deal with could be reduced to a question of value! To put it politely, we have no memory of a collective outcry from scandalised scientists, publicly denouncing one of their colleagues for indulging in such pretences.

But slow science is not – emphatically not – about the goose becoming an omniscient intelligence, able to envisage the consequences of the innovations her science makes possible. Rather, it coincides with the seemingly modest definition given by Whitehead of what universities should foster: rational thought and civilised modes of appreciation. Rational thought would mean being actively lucid about what is actually known, avoiding any confusion between the questions that can be answered in a purified or constrained environment and those that will inevitably arise in the wider and messier environment. A civilised mode of appreciation would imply never identifying what is well-controlled and clean with some truth that transcends the mess. What is messy from the point of view of fast science is nothing other than the irreducible and always embedded interplay of processes, practices, experiences, and ways of knowing and valuing that makes up our common world.

This may be the challenge that slow science should answer, enabling scientists to accept that what is messy is not defective but simply that which we have to learn to live in and think with. The symbiosis of fast science and industry has privileged disembedded knowledge and disembedding strategies abstracted from the messy complications of this world. But in ignoring messiness, and dreaming of its eradication, we discover that we have messed up our world. So I would characterise slow science as the demanding operation that would *reclaim* the art of dealing with, and learning from, what scientists too often consider messy, that is, what escapes general, so-called objective, categories.

A Plea for Slow Science

The term 'reclaiming', as used by US activists, refers to healing operations that would reappropriate what we have been separated from, recovering or reinventing what that separation has destroyed. Reclaiming always begins by accepting that we are sick rather than guilty, and understanding how our environment makes us sick. From this perspective, we might consider the way in which our universities, once so proud of their autonomy, have in the name of the market accepted the imperative of competition and benchmarking evaluation. Likewise, the way in which researchers have accepted without too much resistance the redefinition of research by the knowledge economy. Whatever explanations we can offer, they all testify to the deep vulnerability of what we were once so proud of – the arrangement that promoted fast, disembedded science as a model for scientific research made us too sick to defend it. Playing the goose, researchers accepted a role requiring them to ignore the fact that conquering, destroying and blindly objectifying never had a need for reliable knowledge. Now, however, they understand that competition is generally indifferent to achievements such as the collective production of reliable knowledge; what it requires instead is 'flexibility': scientists who accept that the knowledge they produce is good enough if it leads to patents and satisfies stakeholders.

It may well be that if we had to tell the tale of how scientists and academics were unable to defend the conditions that allow them to exist, we would have to relate how they were finally the victims of the lie that made them modern, allowing them to claim a general authority while the specificity of their practice receded into the background.

Reclaiming operations are never easy. If reclaiming scientific research means re-embedding the sciences in a messy world, it is not only a question of accepting this world as such, but of positively appreciating it, of learning how to foster and strengthen, in Whitehead's words, 'the habits of concrete appreciation of the individual facts in their full interplay of emergent values'.[8] This, as I have already emphasised, does not entail avoiding specialisation and abstraction, which have an obvious value of their own. But concrete appreciation does not just mean abstaining from treating as a mere remainder whatever our abstractions are abstracted from, or abstaining from judging it away. We also need to learn how to actively situate our abstractions in what Whitehead calls the interplay of emerging values. Reclaiming is never only a matter of goodwill, of the kiss of peace turning the disappointing frog into a nice, polite and constructive prince. Learning is needed to get interested in the frog itself, that is, in the mess in which everyone, scientists included, are participants.

Here again we touch upon the radically asymmetrical knowledge developed under the model of fast science. We know a lot about developing material, and so-called immaterial, technologies, but when it comes to much older techniques – the kind needed when people are divided on an issue, and have to learn from each other through their disagreements – we are not very good at all, having lost what we once knew and what other peoples would call civilisation. Just think of the technology of what is becoming a communication imperative, the PowerPoint presentation, and the way it enables one to makes one's point in a striking, authoritative and schematised manner. In 'bullets', no less (just listen to this word. . .).

A Plea for Slow Science

Think also of the boredom we are all so used to, silently and patiently half-listening to a dear colleague speaking for an hour. We have our departments of psychology, social psychology, pedagogy and so on, but we have not learned even a fraction of what activists engaged in reclaiming operations have to learn when they want to work together with others without asserting their authority. They have indeed learned to consider each meeting to be what, following Whitehead, I would call an 'individual fact', depending on the interplay of emerging values; values that can emerge only because the participants have learned how to allow the issue at the heart of their meeting the power to matter, the power to connect everyone present.

Producing knowledge about such individual facts no doubt demands an approach that will not conform to the model of fast science. Moments at which values emerge cannot be disembedded and submitted to general categories; for instance, the moment when someone feels transformed by having understood someone else's perspective; or the gathering that discovers the transformative power of its participants thinking together; or the experience that something which until now appeared insignificant may indeed matter. Such moments have been treated superficially, with inappropriate categories derived from the imperative of reproducibility. They have been judged unfit for knowledge, or worse, relegated to the irrational, and so deemed unworthy of our attention. But it may well be that the approach they need is just a bit different, that what we need to learn is not how to define them, but rather how to foster them. We need to find out what supports and sustains them, and what thwarts or poisons them: to gain something like the slow knowledge of the gardener as opposed to the fast knowledge of 'rationalised' industrial agriculture. In this respect, the kind of knowledge produced in our universities is indeed radically lacking balance, and we are all paying the price for it.

Again, reclaiming means first of all recognising that we are sick and need to heal. Slow science does not provide a ready-made answer; it is not a pill. It is the name for a movement in which many paths to recovery might come together. As for us academics, what about introducing slow meetings, that is, meetings organised in such a way that participation is not only formal? What about slow talks, not just inviting people one really wishes to hear, but reading and discussing beforehand so that the meeting is not reduced to the ritual of attending a prepared lecture that ends with a few banal questions? What about demanding that when colleagues speak or write about issues that are beyond their field of expertise, they present the information, learning and collaborations that have allowed them to do so? What about ensuring, when expertise is needed on an issue of common concern, that co-experts are present and able to represent effectively the many dimensions relevant to the issue? From the point of view of fast scientists, all these proposals have a common defect. They all involve wasting time, or worse, breaking with the symbiotic relation that binds 'true progress' to industrial innovation.

These are only suggestions, and I must admit that I have spent much more time telling you about fast science than about what slow science would be. Accompanying those who today insist that 'another science is possible', my job, as a philosopher, is to try to activate the imagination, which involves going beyond the question of the present mobilisation of research called the knowledge economy to examine the consequences of the older mobilisation. The powerful hold of these consequences on our imaginative resources has to be challenged.

I have tried to confront what has been called 'autonomy', seeing it as a poisonous gift. The name of the poison is progress, mobilisation for the advancement of knowledge as an end in itself, and its consequence is the extraordinary contrast between the imaginative, demanding cooperation between colleagues for whom reliability is the primordial value, and the easy, arrogant way in which those same colleagues dismiss or ignore the world reduced to a field of operation for rational progress.

Challenging mobilisation – which divorces scientists from their power to think, imagine and connect, which defines whatever would slow them down as *necessarily* secondary since what would be slowed down is progress – entails rethinking and reinventing scientific institutions. But I want now to approach the question from another angle, not pre-empting this reinvention, which is not my task as a philosopher, but activating another complementary imagination, which concerns those academic fields without any golden eggs, namely, the humanities.

Indeed, I have heard it said a bit too often that what the golden-egg scientists lack is reflexivity, specifically that critical reflexivity cultivated by the humanities. I have even heard it said that if the humanities are today drastically underfunded it is because this critical reflexivity must be kept at bay, since it poses a threat to mobilisation. My claim, however, is that this reflexivity may also have to be reclaimed as part of the problem rather than the solution, at least in so far as it also defines itself as something that 'others' are lacking, thereby ensuring the humanities' self-proclaimed privileged standpoint: they believe, but we know better; and even better and better with each new theoretical turn.

My position is not to be confused with an acritical one.[9] But I certainly mean to give voice to my deep frustration with the quasi-constitutive relation between critical reflexivity and suspicion, wherein debunking or deconstructing appear as achievements in themselves. This speaks to me of a mobilisation of its own kind, implying that a distance is to be maintained from what others present as really mattering to them.

Whitehead, as I quoted him above, defined the task of the university as *the creation of the future, so far as rational thought, and civilised modes of appreciation, can affect the issue*. Critical reflexivity, to put it in a nutshell, does not seem to me to be engaged by the question of how its own interventions are liable to 'affect the issue'. Indeed, it often seems to be an attempt to compel others – for example those raising issues concerning the creation of a future worth living – to recognise that they are one or many theoretical turns too late. Is not Vandana Shiva's struggle against the patenting or industrialisation of life ignoring the anti-essentialist turn? Nevertheless, I have noted that nowadays the frightening question of climate change has become a popular topic for critical thinkers, under the theme of the 'Anthropocene'. Many rival theoretical turns are in gestation, hunting down new scapegoats, including any colleagues who can be associated with 'anthropocentrism' for having ignored the theoretical challenge of dealing with our species as a 'geological force'. It may well be that such critical thinkers will find many activists' environmental, political and social struggles to be irredeemably 'anthropocentric'.

Reclaiming rational thought from mobilisation, and reclaiming civilised modes of appreciation from their temptation to contrast themselves with others who need enlightenment (whatever light an academic field claims to provide), are clearly not enough. We also have to reclaim the unknown that figures in Whitehead's definition: 'so far as [what we thus reclaim] can affect the issue', that is, can affect other struggles aimed at the creation of a future worth living. This, I would argue, is not a matter of reflexivity. It rather demands what I would call an 'ecology of partial connections', which requires learning from others, being transformed by what is learned, and acknowledging our debt to this transformative experience as we explore its problematising impacts in our own terms.

Making partial connections means first of all accepting being situated. Reclaiming operations, whether conducted by activists, academics, Indian peasants, feminists, or others, are always particular and partial because they are always situated, starting at the very point where we have been humiliated, that is, separated from our power to think, feel, imagine and act. And this is the very reason why the participants need each other and may connect with each other; or rather, need to learn *how* to connect with each other in order to learn and draw new consequences from each other's experience.

This is why, quoting Deleuze and Guattari's *A Thousand Plateaus*, I would say that reclaiming operations speak to us of 'an ambulant people of relayers, rather than a model society'.[10] Referring to William James, I would say that their logic is that of the making of a pluriverse, or, in Mario Blaser's terms, of the weaving of what will always be more than one but less than many.

The test here may well be whether we can reclaim, for those ideas that make us feel and think, the capacity to 'add' something to reality, rather than considering ideas and knowledge in terms of truth, explanation or objectivity. Relaying is never 'reflecting on', but always 'adding to', and thus communicating with what William James defined as the 'great question' associated with a pluriverse in the making: does what we relay 'with our additions, *rise or fall in value*? Are the additions *worthy* or *unworthy*?'[11]

This is a testing question indeed: demanding, as Haraway expresses it, that one consent to 'responsibility' in her sense of the term; accepting that what we add makes a difference to the world and becoming able to answer for the manner of this difference. How, in so doing, do we cast our lot for some ways of life and not others? It should be obvious that casting our lot does not exclude formulating matters of critical concern, but that concern must be such that it is liable to be shared with the people concerned, liable to add new dimensions to the issue they struggle for. And it must thus exhibit what one has learned from them, not herald the general academic concern, which is to create the distance that authorises us as academics.

Relaying is just an example. Going beyond this, I am convinced that reclaiming, for us academics, requires that we collectively learn how to think with James' question: what do our ideas add to what they intervene (or prey) upon? Far from struggling to retain our ancient privileges, we should dare to think with the possibility that we are able to make worthy additions to the weaving of situations that will enable resistance against the coming barbarism. And this may well be the most demanding version of what I called, with James, a genuine option, the challenge to consent or to evade. I described Whitehead's definition of the task of the university as being exposed to sniggering. Here we have to face and feel the snigger inside us, the sad little voice that whispers, 'who do you think you are?' And this is a voice that all too easily takes on the accent of critical reflexivity.

James' question is a test, and consenting to it means first of all taking the question seriously while knowing that no theory will dictate or authenticate the answer, and that it is nobody's job to do so. The worth of an addition, or even the possibility of assigning any value to an addition as such, is not, however, a matter of blind faith. And the point is not to silence the critical voice with some resounding Obamian 'Yes we can!' Consenting to the test means first of all measuring how much we have to learn in order to escape this infernal alternative: either feeling authorised or relying on blind faith.

Activists may indeed help us. I am thinking here, for instance, of the reclaiming operations of neo-pagan activists, and the rituals they experiment with in order to become able to do what they call 'the work of the goddess'. But we can also think of the Quakers' rituals. The Quakers did not quake before their God, but before the danger of silencing the experience that would disclose what was being asked of them in a particular situation, before the danger of answering that situation in terms of predetermined beliefs and convictions. The crucial point in both cases is not, it seems to me, the belief in some supernatural inspiration we might feel free to snigger at. The point is the efficacy of the ritual, an aesthetic one, enhancing what Whitehead called 'the concrete appreciation of the individual facts in their full interplay of emergent values'; or the appreciation of this, always this, concrete situation accompanied by the halo of what may become possible.

We may understand this efficacy in terms of what Deleuze and Guattari called an 'assemblage' [*agencement*], recalling that for them the manner of our thinking and feeling existence is our very participation in assemblages. The reclaiming witches' ritual chant – 'She changes everything She touches, and everything She touches changes' – could surely be commented on in terms of assemblages crafted to resist the dismembering attribution of agency. Does change belong to the goddess as 'agent' or to the one who changes when touched? But the first efficacy of the refrain is in the 'She touches'. Resisting dismemberment is not conceptual. It is part of an experience which affirms that the power of changing is NOT to be attributed to our own selves, nor to be reduced to something 'natural' or 'cultural'. It is part of an experience which honours change as a creation. Moreover, the point is not to comment. The refrain must be chanted; it is part and parcel of the practice of worship.

The point is thus not to theorise assemblages, but to accept that we ourselves are part of academic assemblages which induce and enable us to critically comment and dissect. Taking seriously William James' question may well demand that we learn to live without the protection of such assemblages and to craft different ones: luring assemblages, luring us towards what Whitehead called concrete appreciation. As an act of defiance it may well be that we should, when speaking of the efficacy of such assemblages, dare to use the word the reclaiming witches themselves use: magic.

But we who are not witches do not have to mimic their craft. What they explore is not a speedway to be enthusiastically rushed into, like one more of those famous academic turns. Whatever way we may reclaim the capacity to honour change, it must resist the pressure *inside* academia: that of our dear colleagues who will object that we are not being objective or critical enough, or of journals that insist on the need to respect their norms, the need to begin by expounding 'Materials and Methods' (or the Literature Review!). I would thus claim that if we academics wish to reclaim our practices as worthy, we also need to become reclaiming activists *in our own way*, inventing our own ways of answering the barbarism that gains ground every time we bow down before necessity, including the necessity of either accepting the rules of the game or being excluded from it.

Again, recognising that we are infected and may be spreading the infection is not a matter of guilt to be atoned for, but of learning how to create means of protection. We have to learn, as the witches did, how to cast circles that protect us from our insalubrious, infectious milieu without isolating us from the work to be done, from the concrete situations that need to be confronted. Our pragmatic and empirical concern would then require cultivating, together with those we trust, an informed art of disloyalty, the art of discreetly dismantling academic habits, of confusing the gaze of the inquisitors, of regenerating ways of honouring whatever it is that makes us think and feel and imagine.

As I have emphasised, each reclaiming operation is particular. That is, each has to invent its own means, to create its own interstices, its own ways of protecting itself and of making others feel that resistance is possible. This may be what we should concoct with trusted colleagues, and teach to our students, or those students we trust. It is also, by the way, what resistance movements on the ground learned to do during the Second World War in Europe. This, at the very least, is the kind of tale we should be able to tell to the children born this century when they ask, 'You knew, what did you do?'

6

Cosmopolitics: Civilising Modern Practices

The title given to this chapter foregrounds a word that is rather alluring and a bit mysterious: 'cosmopolitics'.[1] But another word is absent because the organisers of the conference at which it was first presented feared it was liable to create an impression of *déjà vu*, or to lead to misunderstanding.[2] The absent word is a name: Gaia. And yet it is with Gaia that I would like to begin, because it is her intrusion that places me in my current position. She forces me to evoke a possibility that could be rejected twice, and quite rightly. The very idea of 'civilising modern practices' (which, in earlier chapters, I have associated with 'slowing them down'), will be rejected by those who hold that these practices are synonymous with civilisation, bearers of a future in which the whole of humanity will be liberated from the transcendences that divide it and set it at odds with itself. But it will be equally rejected by those who identify these practices with instruments of domination and predation, and for whom the very idea of their possible civilisation is not only an empty idea, but a suspect one. By supposing that they can be presented as 'reformable' doesn't it thus 'relativise' their crimes? Obviously, I wouldn't dream of reconciling these two contradictory positions between which it is so hard to choose. I'd rather create a space for a possible reformulation of their seemingly irreconcilable standpoints. A pipe dream, one might say. But here let me echo the call that was part of feminism's strength: 'things really *could be different!*' And today this call is resonating at the edge of the abyss. To name Gaia is to name a future that could well 'reconcile' our contradictions in reference to a long superseded past, to the times when it was still possible to have debates about 'civilisation'. Barbarism, which is coming, would rule supreme.

So, let's begin with this name, Gaia. It could well encapsulate a paradox of our present times if it can be associated with the fear of creating an impression of *déjà vu*. Whatever meaning we give to this name, it should rather be associated with, or coloured by, a feeling of '*jamais vu*': what cannot be really envisaged – an 'inconvenient truth' indeed, a truth whose radical novelty must be emphasised again and again. At least for the 'we' who have endorsed the 'great divide' with 'peoples' on one side, defined by the way they project their beliefs onto nature, and on the other side a 'we' which is more like a 'one', the anonymous 'one' that 'now knows' in a way destined to finally bring the whole of humanity together. The time is over when this 'we' could think it was free to discuss whether the Earth should be defined as the totality of the resources available for our use, or should be protected. 'We' face a devastating power suddenly intruding into the stories we tell about ourselves but cannot realise, or make real, what is happening.

Then the *déjà vu* may well tell us about the way this knowledge is backgrounded, becoming a weary, 'Yes, I know.' Much more urgent crises are mobilising our attention. But the intrusion of Gaia is not a crisis, in the sense of a transitional period that would allow us to envisage the post-crisis. She will be a permanent part of our future, raising the question: will this future be worth living? As for the fear of misunderstanding, it is certainly elicited by the fact that I gave a name – Gaia, as if it were a person – to what scientists are figuring out to be a complex assemblage of natural processes. Is it a simple metaphor, or am I one of those who 'believe' that the Earth is a being endowed with intentions, if not a consciousness?

Neither one nor the other. Naming is a pragmatic process; its truth depends on its effects. Climate change, and all the other processes that are poisoning life on this Earth, and which have their common origin in what is called development, certainly concern all those who live here, from fish to people. But naming Gaia is an operation addressed to 'us', that tries to arouse an 'us' who would no longer take itself for the anonymous 'one'. We are those who are proud of having defined 'nature' in terms of processes that, together, constitute the scene for primordially human histories – we are those who cannot deny their responsibility for the intrusion of Gaia – and then we are the ones who have created ways of understanding and anticipating some of her effects. This is a new type of division, so to speak, but very different from the first, because it changes the meaning of the word responsibility. We are no longer in charge of the responsibility of showing other people the way to become members of the great 'One' which will henceforth 'know'. We stand before them, as responsible.

James Lovelock chose the name Gaia to characterise this being who is now scrutinised with all the power of the world's centres of scientific observational instrumentation and calculation. Certainly, and unfortunately for us, Lovelock may have been wrong when he proposed that Gaia was, like a healthy organism, gifted with self-stabilising properties. We know only too well, now, that the global result of the complex, non-linear couplings between processes which compose her, and which used to sustain what we have so long taken for granted, was never stable, only metastable, subject to brutal, global mutation. But Lovelock was right to propose that we learn to address this assemblage of processes as an individual being, because the way it answers to perturbations entails a complex and individualised processual coherence, irreducible to a simple sum of modifications. As such, Gaia is questioning us, we who have unleashed a threat to everything we took for granted. And who can predict the difference between the catastrophe of a four-degree increase in average temperature and the cataclysm of a six-degree increase?

Thus, naming this being Gaia is not giving another name to the Earth. Neither is Gaia to be confused with the nourishing land so many peoples care for; nor with the Mother whose primordial rights some demand we recognise and respect. She doesn't contradict these other figures, nor is she their rival. She is adding a further figuration which is specifically relevant for us who belong to a history that has relegated these other figures to the register of 'purely cultural' beliefs.

But the name of Gaia is also the name of a very ancient divinity, a Greek divinity much older than the anthropomorphic gods and goddesses of the Greek cities. It may be that she was a mother figure, but she was not a nice, loving mother – rather more of an awesome one, who should not be offended. She was also remarkably indifferent, with no particular interest in the fate of her offspring. This ancient Gaia corresponds very well to what I name Gaia today: 'the one who intrudes'. Her intrusion is not an act of justice or punishment, because it is not aimed at those who have offended her; rather, it puts a question mark over the future of all inhabitants of the Earth, with the probable exception of the innumerable populations of micro-organisms which, for billions of years, have been the effective co-authors of her ongoing existence. Gaia is this figure of the many-figured Earth which demands neither love nor protection, but the kind of attention to be paid to a prickly powerful being.

I had to begin with Gaia in order to situate my approach, which I would characterise as inseparably constructivist, pragmatist and speculative. The point is not to add a touch of mystery to the intricate inter-coupling of purely material processes that scientists try to decipher. Gaia, as an implacable, unintentional power, blindly answering to the reckless character of what we call progress, is without mystery. Naming her is rather giving a name to the novelty of the event, the irruption of a new kind of transcendence which must be acknowledged by those who equated human emancipation with the denial of any transcendence. Gaia, the one who intrudes, the one whose patience can no longer be taken for granted, is thus not what should unite all the peoples of the Earth. She is the one specifically questioning the fables and refrains of modern history. There is only one real mystery at stake here. It is the answer that we – meaning those who belong to this modern history – may be able to create as we face the consequences of what we have unleashed.

Ours is a time of confusion, anxiety and perplexity. The powers that be seem to have chosen – but did they even choose? – to carry on as if the future had to manage for itself. Their only answer would appear to be that we hold our course, continue to struggle for growth and competition, and trust that some technological fix together with a 'green' capitalism will deal with Gaia. I will not comment on this at length here. I will just emphasise that from the perspective of capitalist logic, the intrusion of Gaia indeed offers new and interesting possibilities, that is, a source of multiple new opportunities to be exploited. But I can only wonder how anybody can really hope that this opportunistic logic will save us from social and ecological disaster. Such a hope is fuelled, rather, by despair: since it is impossible to do otherwise, we *have* to put our trust in capitalism.

Such a blind hope, however, is a quite real temptation, as it allows people to go on living and thinking as usual in a situation in which nothing we are able to envisage seems equal to the challenge. Changing course at a planetary level is in itself a daunting perspective, but it is specifically so today, when what prevails at every level is the imperative of competition, that is, the economic war of one against all. This is why some of those in government who don't believe in 'green capitalism' may conclude that it is better to wait for the time when action will be forced upon us, trusting that we will then find a solution.

This 'wait and see' trust in the pedagogical and mobilising effect of some future catastrophe, forcing a general change of orientation, seems to me awfully misplaced. I rather fear that when the time for mobilisation comes, they will insist that we submit to the quite unpalatable consequences of what will suddenly appear as absolute necessities. The exploitation of tar sand and the expansion of fracking extraction – made necessary, it is said, by the decline of conventional oil production – are just gentle forerunners of what awaits us both ecologically and socially.

No guarantee

As William James often argued, ours is an unfinished world, and action in this world must be divorced from certainty and the demand for guarantees.[3] However, he stressed, we know that what we do, or do not do, the way we consent to the fight or give it up, are part of the making of the future. The Gaia intrusion situates us in a genuine Jamesian option. Trust in an uncertain, indeed improbable, future worth living may seem foolish, but there is no way of avoiding the option because there is nowhere to stand outside of the alternative of either consenting to or refusing the challenge as it is addressed to each of us.

The Jamesian option to consent to the fight does not correspond to a general call for action, even if it may indeed mean joining in street protests and other kinds of legal or not-so-legal actions. It rather demands that we allow ourselves to feel the challenge as addressed to us, as academics, rather than to people in general. There may be people who trust that we, and the students we train, are actively concerned by the part we may be able to play in the creation of the future. When we experience being situated by this trust, we may well feel that the future has already begun. Instead of putting ourselves in the position of our children or our children's children, we might envisage today's answer to our students if they were to ask us: 'What are you doing with what you know? How is it changing your matters of concern?'

If such a question were asked, the answer might have to be that our thinking, imagining, envisaging and proposing are mobilised elsewhere. We might very well know about Gaia, but we hope that the future does not demand that we play a part, however small, because we are really too busy satisfying the relentless demands to which we now have to conform in order to survive. I am not even speaking here about the knowledge economy and the imperative to produce knowledge of interest to the competitive war-games of the corporate world. As we know, even academic fields that produce no patents have now been submitted to the general imperative of benchmark evaluation, having to accept the judgement of an academic market ruled by competition. In sum, whatever the questions the intrusion of Gaia impose on us, it may well be that our research institutions are today quite badly equipped to formulate, or even envisage them.

We also know that the same disempowering processes are at work everywhere. Everywhere there is a proliferation of similar cuts, amputations of our capacity to envisage, feel, think and imagine. If today's struggle must be one we can all agree to, in James' sense of the term, then it might well be the struggle for reclaiming this capacity, or even the capacity to envisage the possibility of reclaiming it. However, one never reclaims in general. Reclaiming operations are initiated at the leading edge of the cut, where each practice has been humiliated, separated from its power to make practitioners think and envisage. My trust is in the plurality of reclaiming operations and the ways in which they may connect, weaving relations with and learning from each other.

Turning now to those practices I would call modern – since one way or another they have defined themselves in terms of the conquest of knowledge and the mission to civilise others – I know that some critical academics might feel unconcerned at the idea of reclaiming and protest, since they no longer endorse this conquering and missionary enterprise. But it is not sufficient simply to disavow the ideas that have blessed that enterprise. What may well remain is irony, perplexity and guilt, in the retreat to purely academic and inconsequential postmodern games.

If our practices have to play a part in reclaiming the capacity to answer to the consequences of the intrusion of Gaia, I propose that they don't just have to give up the idea of a purely human history of progress and conquest, which is precisely what this intrusion challenges. They also have to reclaim a different, positive, definition of themselves and of civilisation, in order to regain relevance and become capable of weaving relations with different peoples and natures.

As you can tell, I am not equal even to our academic situation, as I have characterised it, not even addressing the question of reclaiming what we have effectively surrendered. I am speaking as a philosopher, and more precisely, as a European philosopher, still practising philosophy in a way that has already been mostly destroyed in North America: taking ideas and their adventure seriously. I see my proposal as a little derisory, as will be the case for any particular reclaiming operation. But I do not see it as pointless, because ideas have an efficacy of their own: to poison or to activate, to close down or to open up possibilities.

Philosophical ideas were certainly active in the modern enterprise of civilising conquest. They were mobilised in particular in order to turn modern science into a general and emancipatory model of objectivity, rationality and universality, which, as such, authorised understanding the ways of being and knowing of other peoples as a question of cultural diversity only. It may be because I learned to become a philosopher in close contact with physicists that I felt that this model was a lie. Indeed, these physicists were engaged in an adventure, passionately trying to construct their own questions, to answer problems that were problems of their field's own making, and not at all participating in some consensual advancement of knowledge.

This is why my own contribution to the reclaiming operations we need – a contribution which, I hope, can be connected with others – stems from a double trust: trust in the adventure of ideas, and here, centrally, of the idea of civilisation, which 'really *could be different!*'; and trust that scientists, or at least scientists committed to their science as a very particular selective and demanding practice, may become able to present themselves as such; that is, to reclaim their practice from the lie that has been at work since the origin, since Galileo heralded the event which we now identify as the birth of modern science.

To deal briefly with this last point, we can recognise

Galileo as the discoverer of the possibility of what may well be called an event. For the first time in human history a phenomenon, the frictionless fall of heavy bodies, had been given the power to act as a reliable witness, authorising a particular interpretation against other possible ones. But Galileo presented what he had achieved in a way that backgrounded its selective, highly demanding character, its irreducibility to any free generalisation. He was, in a way, the first 'epistemologist', recruiting concepts of philosophical origin in order to present his achievement as initiating and illustrating a general method aimed at the production of valid knowledge grounded on observable facts. Thus, on the one hand Galileo was the initiator of a collective adventure that unites 'colleagues' in thinking passionately in terms of possible experimental achievements, colleagues who share the need to verify that a claimed reliable witness is able to resist their objections and to force their agreement, because their own future work will depend on such a witness and the new possibilities it opens up. On the other hand, he was the first to promote the general, unilateral authority of science, conquering the world, defining what really matters and what are mere illusory beliefs, thus giving his blessing to the destruction of innumerable other ways of relating, knowing, feeling and interpreting.[4]

The power of modernisation has mobilised the authority of science at least as much as the possibilities opened up by its experimental achievements. Blindly objectifying never needed reliable knowledge. And today, as they have become tools of the knowledge economy, we may say that scientists are the victims of the lie that made them modern, masking the strange specificity of their practice. For this is a strange practice, indeed, which Galileo initiated. It may be characterised as depending on a very particular 'enrolment' of phenomena. Phenomena are invited to accept the role of what we might call 'partners' in a very unusual and entangled relation. Indeed, they not only have to answer questions but also, and first and foremost, answer them in a way that verifies the relevance of the question itself.

We can only dream of another story, in which the unifying thread of what we call Science would have been the demanding, specific character of scientific achievement – the commitment to create situations that confer on what scientists address the power to make a crucial difference in regard to the *value* of their questions. If relevance rather than authority or objectivity had been the name of the game, the sciences would have meant adventure, not conquest. Given what the experimental achievement both demands and presupposes, nobody would then have thought of it as a model to be extended. How indeed can one extend a practice that requires the disembedding of what has to be recruited as a reliable witness, and its redefiniton in the terms of the question it should answer, thus presupposing the intrinsic indifference of the prospective witness to the meaning of this question? Instead of a general ideal of objectivity, a positive, radical, plurality of sciences would have been generated, each scientific practice answering the challenge of relevance associated with its own field.

As a philosopher, I have a vital need for such a dream, for such a counter-factual story, in order to mark the difference between post-modern critical deconstruction and a dissolving operation – the equivalent of what chemists do when they use acid to dissolve amalgamated mixtures into chemically active products. I do not wish to deconstruct what has been called Reason, Objectivity or the Advance of Knowledge in order to uncover, for instance, the conquering machine they conceal. Indeed, such a deconstruction, however legitimate, might justify the conclusion that the knowledge economy is only destroying scientists' illusions, which would make it impossible to acknowledge their outrage, despair and mounting cynicism, or to address them as potential participants in any reclaiming operation. Thus, even if it is factually justified, deconstruction fails from a pragmatic, speculative point of view: from the point of view of its effects, it leaves us with a more desolate, empty world.

On the other hand, dissolving is not to be confused with the struggle against alienation, with freeing innocent, adventurous scientists from the powers that have subjugated them. Scientists were never innocent; they actively took part in the ongoing construction of an asymmetric boundary that would protect their autonomy and resist intruders, while allowing them the freedom to leave their protected spaces in order to participate in the redefinition of our worlds. But, as Donna Haraway insists, non-innocence is something our practices, whether modern or so-called traditional, all share. The question of innocence and guilt should be left to the judges. What matters is rather the possibility of creating relevant modes of togetherness between practices, both scientific and non-scientific; finding relevant ways of thinking together.

And here both critical deconstruction and the knowledge economy have been disastrous. The first provoked the science wars, leading furious scientists to mobilise as the defenders of Reason under attack. The second entails the production of scientists unable to account for their choices about what matters and what does not, since those choices will be defined by the interests they serve. Again, I am following William James' pragmatism here, giving primordial importance to the making of relations, the construction of what he would call a pluriverse, even identifying the relation-making capacity as synonymous with civilisation.

Such a capacity is a testing one. It puts a constraint on the way one presents oneself, and indeed, thinks of oneself. Humans hardly present themselves to other humans as creatures gifted with opposable thumbs, however crucial this feature might be, but a scientist may well think of her practice as being objective or rational and present herself in such terms. This is an insulting thing for her to do, as it implies that these distinctive characteristics are lacking in the one she is addressing. But civilisation understood as the cultivation of an art of relation-making also precludes whatever would turn relation-making into the normal outcome of something more general, such as Habermas' idea of communicative rationality. Relation-making does not consist simply in the recognition that we are related; it is an achievement. It implies the risk of failure, the hesitation between peace and war.

From that point of view, the scientist can again be taken as an example. The experimental achievement is a case, a very specific case, of relation-making between passionate human beings and what might verify the relevance of their questions. Such achievements may be seen as the creation of bridges between heterogeneous beings gifted with radically divergent ways of behaving, bridges that open up new possibilities of action and passion on both sides. Those scientists for whom this kind of relation-making practice matters, that is, who are not serving a so-called scientific method, know very well that it would be destroyed if the questions to be answered were imposed upon them. They had anticipated this possibility ever since the second half of the nineteenth century, arguing that the subjugation of research to non-scientific interests would be like killing the goose that laid the golden eggs. The goose demands to be left alone; she is not accountable for the use of her eggs, she just demands that her own relation-making, both with her colleagues and with what matters for herself and her colleagues, be respected.

Of course many scientists have been, and are now more than ever, passionately engaged in the creation of relations with industrial and State interests. Here the valorisation of the eggs prevails over the kind of concern that would characterise civilised science, which would have publicly presented its reliability as depending on the social fabric of competent colleagues interested in testing and contesting its results, results that are thus also situated by this social fabric.

Civilised scientists would be the first to affirm that both the reliability of their results and the competence of their objecting colleagues are relative to experimentally purified, well-controlled laboratory experiments, which require ignoring what may be important factors outside the laboratory. They would thus acknowledge that whatever they achieve may well lose this specific reliability when it leaves the network of research laboratories. The only way to regain reliability would then be to weave new relations proper to each new environment, and to welcome new objections – no longer just from colleagues, but also from those for whom this new environment is a matter of active concern.

Again, this story about civilised science has a dreamlike quality, just like the one in which relevance would have been the unifying thread of what we call Science. And, again, the dream is to dissolve the amalgamated mixture that our own history has produced, where only certain questions will be taken into account, while others, identified with a subjective, irrational resistance to progress, will be ignored. We now have to acknowledge the result: up until the advent of the knowledge economy, scientists may well have protected the reliability of scientific claims, but they were active participants in a mode of development that we are now forced to recognise as having been radically unsustainable, and as becoming more so today.

The two dreamlike stories I have outlined here serve to situate the ambition to reclaim that is one aim of what is now called 'political ecology', itself a response to the radical unsustainability which now provokes the intrusion of Gaia. The stories throw light on three features of this political ecology, while pointing out a limitation.

Political ecology

The first feature is that political ecology needs to 'put the sciences into politics', but without reducing them to politics. This requires fully developing, around each issue, the primordial political question: who can talk of what, be the spokesperson of what, represent what, object in the name of what? The invention of the modern experimental demonstration itself can then be understood as a particular answer to this question, an answer specific to the issue of reliability which prevails in the experimental environment. Reclaiming it as such, resisting its hijacking by a general model of objective, rational knowledge, means that a continuation of the political question is needed in each new environment, requiring new spokespersons, and framing new issues.

In order to participate in such political ecological negotiation, as characterised for instance by Bruno Latour in his *Politics of Nature*,[5] researchers would be required to present what they know in a civilised mode, a mode that openly situates this knowledge in relation to the precise questions they are able to answer. They should, in other words, render this knowledge 'politically active', engaging it in a collective assessment of the differences it may eventually make to the formulation of an issue and its envisaged solutions.

The second feature is obvious. A choice has to be made between political ecology and political economy, and more precisely what I called above capitalist logic. I would characterise this logic as intrinsically incapable of being civilised, because what matters for it is not possibilities for relations, but opportunities for exploitation. It can be said that, before finally taking direct control of scientific research, capitalist logic fully exploited the opportunities opened up not only by scientific results but also by scientific claims to general objectivity and rationality. Scientists were offered the possibility of being productive geese, the innocent agents of a development they allowed to be presented as authorised by rationality.

The third and correlative feature, before I come to the limitation, is the need to resist not only the knowledge economy, which is obvious, but also the kind of training scientists receive in modern academic settings, which are dominated by the sharp opposition between questions defined as scientific and those that should be left to politics, or rather to 'ethics' (which has taken the place of politics). Expressions of goodwill and verbal submission on ethical matters will never produce scientists capable of being interested in the objections of all concerned parties in an issue, and able to respect them as they respect their colleagues' objections. This does not mean that scientists should become generalists. But it certainly means that they should cultivate an active, concrete awareness of the very special and demanding character of their knowledge, and the way its reliability depends on the distribution between what they define as mattering and what can be ignored. Acquiring and maintaining such a concrete awareness, as a condition for the capacity to enter into new relations, takes time, and this may be the true challenge here. For scientists educated in modern research institutions, whatever requires slowing down mobilisation amounts to a distraction, a diversion from the scientists' one true mission of advancing knowledge. We thus need the same kind of deep change that slow food movements propose.

I come now to the limitation, and to what I have called cosmopolitics. The latter term came to me as something of a surprise, when I suddenly realised that political ecology itself had to be civilised. I was working on the formulation of what should be demanded of participants assembled around an issue, in order to give that issue the power to get them thinking together; the conclusion I came to was that all participants would have to accept that the meaning of what matters for each of them, or what they are the spokespersons for, is to be determined by the relations woven through this thinking together. But then I realised that what I was formulating were the conditions of a political process as my own tradition had defined it, a process that admits no transcendence.[6]

Civilising politics

The intrusion of Gaia is a danger for all natures and all peoples on the Earth, but it may also legitimise the brutal demand that all peoples acknowledge that they are in the same boat, that they all have to agree to present what they know to each other, but in a way that renders these knowledges 'politically active', liable to political reinvention. What I call cosmopolitics is not the solution to this difficulty, but a name for it, a name calling for the invention of modes of gathering that complicate politics by introducing hesitation. This is what Donna Haraway has now turned into a thought-provoking motto: 'staying with the trouble'.[7]

Cosmopolitics is about resisting the temptation to rush to the conclusion that political ecology is finally the right solution, with which all the peoples of the Earth should agree, or else be excluded on the grounds of fanaticism and irrationality. Politics, even political ecology, has to think of itself in a civilised manner. Cosmopolitics has therefore nothing programmatic about it. It has far more to do with activating a passing tremor of fright in a gathering that may be tempted to think it is sufficient to give every concerned party a voice: 'We are ready to hear your objections, your proposals, your contribution to the issue we are gathered around.' I am a daughter of the world which invented politics, and political ecology situates me as belonging to this world. Cosmopolitics still belongs to this particular world but it doubles up the issue that is to be politically formulated, with an awareness that some formulations may attack the very fabric of other worlds.

Cosmopolitics demands that the political scene be conceived in such a way that collective thinking proceed 'in the presence of' those who belong to these worlds, and who risk otherwise being unheard because they refuse to accept that the meaning of what they are attached to will be determined by the political process. They may be disqualified because, far from contributing to this process, they are hindering an emergent agreement. The cosmos, as alluded to in cosmopolitics, thus intervenes as a way of 'slowing down', of resisting the idea that it must matter for everybody that a correct position be reached, which should be accepted by all those concerned.

We could say that the cosmos, here, acts as an equalising operator, slowing down the political voices mobilised by the agreement to be crafted, imbuing them with the feeling that not every concerned party might have, can have, or wants to have, a political voice. Equalisation is thus distinct from political equivalence, demanding that everybody have the same, equivalent, say about an issue. It rather demands that all concerned parties be present in the mode that makes the decision concrete, that is, as difficult as possible, precluding any shortcut or over-simplification, any *a priori* differentiation between that which counts and that which does not.

The cosmos of cosmopolitics must therefore be distinguished from any particular cosmos or world as a particular tradition may conceive of it, or from something that would transcend all of them. There is no representative of the cosmos as such, no one talks in its name, and it is not a matter of special concern. Its mode of existence is rather reflected in an artificial staging to be invented, the efficacy of which would be to expose, to the fullest extent, the consequences of decisions that are made.

A first aspect of this artificial staging, I would suggest, involves making an active distinction between the figure of the expert and that of the diplomat. I call experts those who give voice to a position that is able to accept the constraints of the political procedure, that is, those who are called to contribute to a relevant decision representing a group that will not be existentially threatened by this decision, whatever it may be and whatever the way their contribution is taken into account by it. The experts' role requires them to present themselves, and to present what they know, in a mode that does not preempt how that knowledge should be taken into account. By contrast, diplomats are there to provide a voice for those whose practice, mode of existence, world, or what is often called identity, may be threatened by a decision: 'If you decide that, you'll destroy us.' The diplomats' role is therefore above all to force experts to think about the possibility that an envisaged course of action may effectively amount to an act of war.

It is important to emphasise that the distribution of diplomats and experts is not an essentialist one. It is issue-based; that is, it reflects the position of each concerned group in relation to the formulation of the issue. Even scientists may need diplomats, because their world can also be destroyed, as indeed it is by the 'knowledge economy'.

However, this distribution of roles may be quite insufficient. Diplomats relate to the possibility of war, and their role entails that those they represent are able, when the diplomats return with a proposal, to organise some form of consultation process about it and decide between agreement and resistance (or war). The practice of consultation, of collectively determining what is or is not acceptable, is a demanding one in itself, which again may easily become a factor of discrimination. What about what I would call the 'weak' parties, those who are unable or unwilling to send diplomats, those who have no spokespersons, nobody to defend them or speak in their name? I would suggest calling them 'victims', because victims need witnesses. It is the witnesses' role to make them 'present'; not arguing in their name but conveying what the issue may mean for them. It is their role to denounce any downplaying of the consequences, any anaesthesia about the price the voiceless ones might have to pay for the political game being played out over their heads.

The presencing of the victims is obviously no guarantee of anything, any more than is the diplomatic intervention. Cosmopolitics has nothing to do with the miracle of decisions that 'make everyone agree'. It rather concerns the demand that decisions be taken in the full and vivid awareness of their consequences. No decision is ever innocent. What is important here is the prohibition against ignoring, forgetting, or, worse still, humiliating. Those who gather around an issue have to know that nothing can erase the debt binding their decision to its eventual victims. As Donna Haraway puts it, speaking about the suffering of animals killed for our benefit, the point should not be to define some of them as having rights, sharing with us the protection afforded by 'Thou shall not kill.' It is rather that we should never take for granted the legitimacy of the sacrifice of any one of them: 'Thou shall not make killable.'[8] Here one can say: 'Thou shall not define as dispensable.'

It may be objected that this is mere science fiction, or speculative fabulation, incapable of helping us in what is our most urgent task of facing up to the challenge associated with the intrusion of Gaia. As I have already remarked, my real concern is about what is already happening now, and will intensify when the urgency is finally acknowledged. I do not know, nobody knows today, if and how we will be able to compose with Gaia, to answer what is not her challenge but the challenge of the intrusion we have unleashed. I am part of a generation that will have disappeared by the time this question is properly engaged with. But my conviction is that we can already sense what is looming, the kind of dire measures which, it will be said, must be accepted because they are the only ones possible, even if they put into question the possibility of lives that are worth living. This conviction situates me as part of a generation that may turn out to be the most hated in human memory. We knew, and just felt guilty. This is what makes me think in terms of resisting and reclaiming, or, in Donna Haraway's term, of regeneration.

One never reclaims or resists in general. My way of resisting and reclaiming may well seem derisory since it deals with ideas. But the power of ideas is not to be downplayed. The idea that we belong to a tradition that is doomed to define other peoples as entertaining mere beliefs, or nature as a mere resource, is a very infectious one, which you meet everywhere. It breeds guilt and poisons our capacity to resist, leading us instead to identify with the capitalist logic that has captured us. As for the idea of cosmopolitics, its efficacy, however speculative, is to activate the possibility of resisting and reclaiming what this capture has systematically attacked or poisoned. The idea is not to transcend the particularity of the so-called modern tradition, but to think with this particularity, to induce the capacity to imagine the possibility that it can be regenerated or civilised – which does not mean universalised. On the contrary, it means thinking with its own specific, dangerous and never innocent ways of weaving relations. It means thinking with the resources – imaginative, scientific and political – that it may be able to activate in order to enable us, perhaps, to think with other peoples and natures.

We do not know if, or how, we will be able to compose with Gaia, but we have no other option than to trust that we can make a difference, however small, a difference that calls for other differences to be made elsewhere. What I have related is just a tale, which, as such, certainly cannot hope to make 'the' difference. But it does call for other tales, for a weaving of regenerative, slightly transgressive imaginations. Such a weaving might indeed make a difference as it brings with it the possibility of sharing and cooperating – which, while certainly not sufficient, is perhaps a necessary condition for reclaiming a future worth living.

Notes

1 Towards a Public Intelligence of the Sciences

1 An earlier version of this chapter appeared in *Alliage*,
   69 (October 2011), pp. 24–34. It was adapted from
   a keynote speech given to the European Science
   Education Research Association (ESERA) meeting
   in Lyon, 2011.

2 *Mot d'ordre* is a concept developed in the philosophical works of Deleuze and Guattari, variously
   translated as 'slogan', 'directive' or 'watchword'.
   In this first instance it is retained so the reader
   can better recognise its philosophical provenance.
   [Trans.]

3 An extract published in *Le Monde*, 22 December
   2004, p. 18.

4 Jean-Marc Lévy-Leblond, *L'Esprit de sel*, Paris:
   Seuil, 1984, p. 97.

5 This experiment was carried out under the auspices
   of the PAI (Pôle d'action interuniversitaire), entitled

'Knowledge Loyalties' ('Les loyautés du savoir'), led by Serge Gutwirth.

6 Naomi Oreskes and Erik M. Conway, *Merchants of Doubt: How a Handful of Scientists Obscured the Truth on Issues from Tobacco Smoke to Global Warming*, New York: Bloomsbury Press, 2010.

7 A reference to the popular French philosopher, Bernard-Henri Lévy, who played a strident and hawkish role in the decision to go war to establish a democracy in Libya.

2 Researchers With the Right Stuff

1 This chapter was first presented at the conference, 'L'*Homo academicus* a-t-il un sexe? L'excellence scientifique en question', which took place at the University of Geneva on 15 October 2009. An earlier version was published as 'L'étoffe du chercheur: une construction genrée?', in Farinaz Fassa and Sabine Kradolfer, eds., *Le Plafond de fer de l'université. Femmes et carrières*, Zurich: Éditions Seismo, 2010, pp. 25–40.

2 Tom Wolfe, *The Right Stuff*, New York: Farrar, Strauss and Giroux, 1979.

3 Luc Boltanski and Laurent Thévenot, *On Justification: Economies of Worth*, trans. Catherine Porter, Princeton: Princeton University Press, 2006.

4 Virginia Woolf, *Three Guineas*, Orlando: Harcourt, 1966 [1938].

5 An allusion to the French movement 'Sauvons la recherche', which since 2004 has organised public demonstrations and petitions. [Trans.]

6 Woolf, *Three Guineas*, p. 63.

7 Woolf, *Three Guineas*, p. 105.

8 Elizabeth Potter, *Gender and Boyle's Law of Gases*, Bloomington and Indianapolis: Indiana University Press, 2001.

9 Donna Haraway, *Modest_Witness@Second_ Millennium. FemaleMan©_ Meets OncoMouse: Feminism and Technoscience*, New York: Routledge, 1997. Haraway knew of Potter's work well before it was published.

10 Robert Musil, *The Man without Qualities*, London: Picador, 2011 [1943], Vol. 1, Ch. 72.

11 In *Women Who Make a Fuss: The Unfaithful Daughters of Virginia Woolf* (Minnesota: Univocal, 2014 [2011]), Vinciane Despret and I proposed – as Woolf's unfaithful daughters who had embarked on university careers despite her warning – a minor treasonous trope: learn to make a fuss, and relate to others who make fuss, even when there is no hope of winning, or scarcely any. . . . Refuse to accept with courage and dignity that which can't be avoided.

12 See http://sciencescitoyennes.org.

13 Donna Haraway, 'Situated Knowledges: The Science Question in Feminism and the Privilege of Partial Perspective', *Feminist Studies*, 14:3 (Autumn, 1988), pp. 575–99.

3 Sciences and Values: How Can we Slow Down?

1 Here I have been inspired by discussions in the context of the 'Groupe d'études constructivistes' at the Free University of Brussels, based on Katrin Sohldju's work, 'Interessierte Milieus oder die experimentelle Konstruktion "überlebender" Organe', in Karin Harrasser et al., eds., *Ambiente. Das Leben und seine Räume*, Vienna: Turia, 2010, pp. 51–64.

2 On this subject, see Foucault's works on governmentality and its instruments. What I call a convention is related to his 'practical ensembles'. Choosing the term 'convention' raises the question of the type of care that the maintenance of a convention demands.

3 John Dewey, *The Public and Its Problems*, New York: Holt, 1927.

4 Bruno Latour, *Reassembling the Social: An Introduction to Actor-Network-Theory*, Oxford: Oxford University Press, 2005, pp. 93–106.

5 The sphere, rolling from a fixed height along an inclined plane, 'should' fall here: for a reconstruction of this experiment conducted in 1608, see Isabelle Stengers, *La Vierge et le Neutrino*, Paris: Les Empêcheurs de penser en rond, 2006.

6 Richard Dawkins' notion of the extended phenotype is characteristic of this point of view. It allows, as the author admits, the translation of any particular history in terms of the same moral: population genetics. Recalling pre-Copernican astronomy, which 'saved' celestial movements – that is, reduced them to interlaced circles, or epicycles – it may be said that Dawkins' notion, which extends the concept of the phenotype (defined as genetically determined in the last instance) to everything that equips the animal in its milieu (spider webs, beaver dams, human books), is an epicycle-producing machine.

7 Edited by Hilary Rose and Steven Rose, *Alas, Poor Darwin: Arguments Against Evolutionary Psychology*, London: Vintage, 2012.

8 See Vinciane Despret, *Penser comme un rat*, Versailles: Éditions Quae, 2009.

9 The issue is just as likely to come up in the law

('How do we want to be judged?'). See Paul De Hert and Serge Gutwirth, 'De seks is hard maar seks (dura sex sed sex). Het arrest K.A. en A.D. tegen België', *Panopticon*, 3 (2005), pp. 1–14.

10 Elinor Ostrom, *Governing the Commons: The Evolution of Institutions for Collective Action*, Cambridge: Cambridge University Press, 2015.

4 Ludwik Fleck, Thomas Kuhn and the Challenge of Slowing Down the Sciences

1 Ludwik Fleck, *Genesis and Development of a Scientific Fact*, Chicago: University of Chicago Press, 1979.

2 Thomas Kuhn, *The Structure of Scientific Revolutions*, Chicago: University of Chicago Press, 1962.

3 Fleck, *Genesis and Development*, p. 101.

4 Fleck, *Genesis and Development*, p. 98.

5 Fleck, *Genesis and Development*, p. 94.

6 Fleck, *Genesis and Development*, p. 78.

7 Fleck, *Genesis and Development*, p. 78.

8 Fleck, *Genesis and Development*, p. 77.

9 See http://slow-science.org/slow-science-manifesto.pdf.

10 See Chapter 5 of this volume.

11 Ludwik Fleck, 'Zur Krise der "Wirklichkeit"', quoted in Johannes Fehr, '"... the art of shaping a democratic reality and being directed by it ..." – Philosophy of Science in Turbulent Times', *Studies in East European Thought*, 64:1–2 (2012), pp. 81–9 (here pp. 84–5).

12 Fleck, 'Zur Krise', quoted in Fehr, '"... the art of shaping a democratic reality ..."', p. 85.

5 'Another Science is Possible!' A Plea for Slow Science

1 Gilles Deleuze and Félix Guattari, *What is Philosophy?*, trans. Graham Burchell and Hugh Tomlinson, London: Verso, 2003, p. 108.

2 See Isabelle Stengers, *In Catastrophic Times: Resisting the Coming Barbarism*, Open Humanities Press/Meson Press, 2015.

3 A. N. Whitehead, *Modes of Thought*, New York: The Free Press, 1968, p. 171.

4 A. N. Whitehead, *Science and the Modern World*, New York: The Free Press, 1968, p. 196.

5 Whitehead, *Science and the Modern World*, p. 197.

6 Whitehead, *Science and the Modern World*, p. 205.

7 Whitehead, *Science and the Modern World*, p. 198.

8 Whitehead, *Science and the Modern World*, p. 246.

9 In 'Experimenting with Refrains: Subjectivity and the Challenge of Escaping Modern Dualism', *Subjectivity*, 22 (2008), pp. 38–59, I proposed a distinction between critique and discrimination, two words with the same etymological root.

10 Gilles Deleuze and Félix Guattari, *A Thousand Plateaus*, trans. Brian Massumi, Minneapolis: University of Minnesota Press, 1987, p. 377.

11 William James, *Pragmatism: A New Name for Some Old Ways of Thinking*, New York: Longman Green and Co., 1907, p. 98.

6 Cosmopolitics: Civilising Modern Practices

1 An earlier version of this chapter was presented on 5 March 2012 at St Mary's University, Halifax, Canada, under the title 'Cosmopolitics: Learning to Think with Sciences, Peoples, Natures'.

2 It is for the same reason that my book *In Catastrophic Times: Resisting the Coming Barbarism*, first published in French in 2009, was not titled 'The Intrusion of Gaia'.

3 William James, *The Will to Believe and Other Essays in Popular Philosophy*, New York: Dover, 1956 [1897].

4 For more on this subject, see my book *La Vierge et le Neutrino*.

5 Bruno Latour, *The Politics of Nature: How to Bring the Sciences into Democracy*, trans. Catherine Porter, Cambridge, MA: Harvard University Press, 2004 [1999].

6 See Isabelle Stengers, 'The Curse of Tolerance' (1997), in *Cosmopolitics II*, trans. Robert Bononno, Minneapolis: University of Minnesota Press, 2011.

7 Donna J. Haraway, *Staying with the Trouble: Making Kin in the Chthulucene*, Durham, NC: Duke University Press, 2016.

8 Donna J. Haraway, *When Species Meet*, Minneapolis: University of Minnesota Press, 2007, p. 80.